Applications of Conscious Innovation in Organizations

Jesus Enrique Portillo Pizana
Tecnológico de Monterrey University, Mexico

Sergio Ortiz Valdes
Tecnológico de Monterrey University, Mexico

Luis Miguel Beristain Hernandez
Tecnológico de Monterrey University, Mexico

A volume in the Advances in
Business Strategy and Competitive
Advantage (ABSCA) Book Series

IGI Global
DISSEMINATOR OF KNOWLEDGE

Published in the United States of America by
IGI Global
Business Science Reference (an imprint of IGI Global)
701 E. Chocolate Avenue
Hershey PA, USA 17033
Tel: 717-533-8845
Fax: 717-533-8661
E-mail: cust@igi-global.com
Web site: http://www.igi-global.com

Library of Congress Cataloging-in-Publication Data

Names: Portillo Pizana, Jesus Enrique, 1968- author. | Ortiz Valdes, Sergio, 1962- author. | Hernandez, Luis Miguel Beristain, 1961- author.
Title: Applications of conscious innovation in organizations / by Jesus Enrique Portillo Pizana, Sergio Ortiz Valdes, and Luis Miguel Beristain Hernandez.
Description: Hershey : Business Science Reference, [2018]
Identifiers: LCCN 2017027628| ISBN 9781522540236 (hardcover) | ISBN 9781522540243 (ebook)
Subjects: LCSH: Organizational change. | Technological innovations--Management.
Classification: LCC HD58.8 .P6735 2018 | DDC 658.4/063--dc23 LC record available at https://lccn.loc.gov/2017027628

This book is published in the IGI Global book series Advances in Business Strategy and Competitive Advantage (ABSCA) (ISSN: 2327-3429; eISSN: 2327-3437)

British Cataloguing in Publication Data
A Cataloguing in Publication record for this book is available from the British Library.

Advances in Business Strategy and Competitive Advantage (ABSCA) Book Series

ISSN:2327-3429
EISSN:2327-3437

Editor-in-Chief: Patricia Ordóñez de Pablos, Universidad de Oviedo, Spain

MISSION

Business entities are constantly seeking new ways through which to gain advantage over their competitors and strengthen their position within the business environment. With competition at an all-time high due to technological advancements allowing for competition on a global scale, firms continue to seek new ways through which to improve and strengthen their business processes, procedures, and profitability.

The **Advances in Business Strategy and Competitive Advantage (ABSCA) Book Series** is a timely series responding to the high demand for state-of-the-art research on how business strategies are created, implemented and re-designed to meet the demands of globalized competitive markets. With a focus on local and global challenges, business opportunities and the needs of society, the **ABSCA** encourages scientific discourse on doing business and managing information technologies for the creation of sustainable competitive advantage.

COVERAGE

- Strategy Performance Management
- Adaptive Enterprise
- Small and Medium Enterprises
- Tacit Knowledge
- Competitive Strategy
- Differentiation Strategy
- Value Chain
- Core Competencies
- Joint Ventures
- Value Creation

IGI Global is currently accepting manuscripts for publication within this series. To submit a proposal for a volume in this series, please contact our Acquisition Editors at Acquisitions@igi-global.com or visit: http://www.igi-global.com/publish/.

The Advances in Business Strategy and Competitive Advantage (ABSCA) Book Series (ISSN 2327-3429) is published by IGI Global, 701 E. Chocolate Avenue, Hershey, PA 17033-1240, USA, www.igi-global.com. This series is composed of titles available for purchase individually; each title is edited to be contextually exclusive from any other title within the series. For pricing and ordering information please visit http://www.igi-global.com/book-series/advances-business-strategy-competitive-advantage/73672. Postmaster: Send all address changes to above address.

Titles in this Series

For a list of additional titles in this series, please visit:
https://www.igi-global.com/book-series/advances-business-strategy-competitive-advantage/73672

Business Strategies and Advanced Techniques for Entrepreneurship 4.0
Jose Manuel Saiz-Alvarez (Tecnológico de Monterrey, Mexico)
Business Science Reference • ©2018 • 231pp • H/C (ISBN: 9781522549789) • US $195.00

Ethics and Decision-Making for Sustainable Business Practices
Ionica Oncioiu (Titu Maiorescu University, Romania)
Business Science Reference • ©2018 • 308pp • H/C (ISBN: 9781522537731) • US $215.00

Handbook of Research on Entrepreneurial Ecosystems and Social Dynamics in a ...
Luísa Cagica Carvalho (Universidade Aberta, Portugal & Universidade de Évora, Portugal)
Business Science Reference • ©2018 • 601pp • H/C (ISBN: 9781522535256) • US $345.00

Start-Up Enterprises and Contemporary Innovation Strategies in the Global Marketplace
Rajagopal (EGADE Business School, Tecnologico de Monterrey, Mexico) and Ramesh Behl (International Management Institute Bhubaneswar, India)
Business Science Reference • ©2018 • 341pp • H/C (ISBN: 9781522548317) • US $235.00

Global Practices in Knowledge Management for Societal and Organizational Development
Neeta Baporikar (Namibia University of Science and Technology, Namibia & University of Pune, India)
Business Science Reference • ©2018 • 475pp • H/C (ISBN: 9781522530091) • US $240.00

Exploring the Global Competitiveness of Agri-Food Sectors and Serbia's Dominant ...
Svetlana Ignjatijević (University of Business Academy in Novi Sad, Serbia) and Drago Cvijanović (University of Kragujevac, Serbia)
Business Science Reference • ©2018 • 193pp • H/C (ISBN: 9781522527626) • US $165.00

For an entire list of titles in this series, please visit:
https://www.igi-global.com/book-series/advances-business-strategy-competitive-advantage/73672

IGI Global
DISSEMINATOR OF KNOWLEDGE

701 East Chocolate Avenue, Hershey, PA 17033, USA
Tel: 717-533-8845 x100 • Fax: 717-533-8661
E-Mail: cust@igi-global.com • www.igi-global.com

Table of Contents

Preface vi

Acknowledgment xiii

Chapter 1
Ignition 1

Chapter 2
The Definition of DNA 43

Chapter 3
Innovation Initiatives 96

Chapter 4
Integrating Human Talent 133

Chapter 5
Prototyping 150

Chapter 6
Validation 179

Chapter 7
Lean Operation 203

Chapter 8
Evolution 247

About the Authors 293

Index 295

Preface

Stragile is an effort to integrate various contributions in terms of innovation and value creation to guide the conscious action of business units in regional areas. It is not an exhaustive review of the different contributions and models proposed in the last decades, in which case the present work is incomplete and unjust, but of a choice of those contributions relevant to the actors of a regional economy.

Stragile represents the coordinated effort of government agencies, business organizations and higher education institutions to renew the productive and competitive base of Ciudad Juarez, Chihuahua, through the revitalization of established companies, the creation of new companies and the attraction of innovative companies to be inserted in the global value chains of constantly renewing industries such as automotive, emerging biomedical or solutions for industry 4.0. The first results have led to the creation of a technological activity center, the Center for Research and Integration of Advanced Technologies, to strengthen and coordinate the different research and development resources of the region, creating an advanced development ecosystem to promote new levels of research and development, productivity, global human talent and products for the world, leaving behind an efficient industrial base for the "maquila industry".

Stragile proposes a method of "nineteen steps" to renew the relevance of companies through product innovation, process and business model, following an evolutionary and dynamic process. Three assumptions of the model are fundamental: the first assumes that the product is the solution to a customer's problem, being able to assume tangible forms, what is known as a product in the current language, but also intangible as the services itself, and, in most cases, a combination of both. The second, that innovation, as proposed by Peter Drucker, is the specific "tool" of entrepreneurship and this, in turn, a function that can be performed by intra-entrepreneurial teams in established companies or entrepreneurs since the Startup.

Finally, if the objective is to provoke conscious innovation, resorting to entrepreneurial techniques is consequently pertinent even for the established company and with history, recognizing the lack of attention to the organizational culture.

The starting point or first step refers to the sensitization of the different actors in the company, in the coordinating organizations and in the leadership groups to adopt a transformation purpose and a shared vision of rupture, both at a regional level and at a local level from each of the companies willing to join the effort to develop vigorous and competitive initiatives. Being aware implies committing to a purpose and the consequent action to make it a reality. Any cooperative system, be it a company or a community, is firstly structured (understood as the formal provision for the coordination of actions with purpose); but it is also culture in terms of the values and attitudes that influence the action with purpose, and finally it is political in that the action without purpose must be persuaded by the leadership and controlled by the position of authority.

The second step is the motivation, those internal and external factors to the cooperative system that inspire the action, generally modeled by a strong leadership capable of integrating high performance teams. A daring project of such magnitude with the participation of different actors requires "large" innovative teams with fresh approaches collaborating in limited time and resources conditions from different backgrounds, skills and experiences. In these cases, the power of a great transforming vision is capable of injecting the enthusiasm and the necessary spirit to undertake disproportionate challenges.

The third step refers to understanding the causes, the conditions and the nature of the present situation, the diagnosis, based on the analysis of megatrends as a frame of reference to identify the opportunities for creating value. Without pretending to embrace the totality of the forces that are transforming society as we know it, particular attention is paid to two inescapable forces: demographic change and technological change. By force to alter social institutions, rethink values, organize and develop new megacities, and the growing occupation of the sustainability of our way of life. The second, in particular what is known as Industry 4.0 from exponential technologies, augurs deep changes in business models, manufacturing and industrial organization as we know them. The integration of the physical world with the virtual, resorting to digitalization, automation and robotics, are causing the transformation. The challenges are huge and real, but the progress of the solutions is also real. From the abundance approach, there are multiple opportunities for people, companies and regional economies, if they are able to understand the forces behind the transformation and the best way to take advantage of them profitably. Internet of Everything, the growing interest in energy consumption, in energy conservation and the ecological footprint, are new demands that offer the possibility of adding productively to the new global value chains.

The next step is to encourage human talent and entrepreneurial mentality. Much has been written about the first, and relevant lifelong learning is necessary in a dynamic and changing environment. On the second, suffice to say that the entrepreneurial mentality is characterized by the desire to perform "something" worthwhile, mobilize the means to put them into action and learn dynamically in the process; is to have the ability to visualize a better future and strive to make it a reality. By nature, the human being is an entrepreneur, otherwise we would not have reached the point where we are; the point is to do it minimizing risk conditions by using the techniques of agile thinking, design thinking and exponential thinking.

The fifth step focuses on the guidance team, the persons in charge of mobilizing the wills for transformation, distinguishing key elements for their proper integration and functioning, such as communication for learning, innovation and entrepreneurship; the verbalization of the purpose of transformation in a shared and inspiring vision; the clarity of direction and action through agile strategic planning, characterized by respect, trust and loyalty in the capacities and integrity of the members to achieve the common goal. The success of a transformation effort depends to a large extent on the quality of the guidance team and the care given to its integration. The credibility of the members for their integrity as persons and their professional quality increases the chances of success of the transformation effort.

We previously referred to the importance of the vision to inject the energy and enthusiasm of all those involved in the transformation projects. The DNA, the sixth step, is not the product of inspiration nor finds its foundation in a belief or destiny. On the contrary, it is the result of diligent diagnosis, the identification of opportunities, the choice of a future and the articulation of actions through the organizational architecture and the resources to reach it. As a process, it might look as strategic planning, but the end of his nature are usually rational, cold and objective, and this is no longer the "unique" answer modern companies are looking for. Developing a vision and answering the question "why" to commit to it, is a matter of giving it a meaning rooted in the aspirations and values of the group, explaining why it is valuable and why it is worth pursuing it, communicated through aspirational images.

Without a transforming vision, it will hardly move the spirit of all those who are indispensable to make it a reality. *Leadership is nothing other than the ability to enable others in the sense of purpose for any task or activity that is required.*

Step 7 is focused on maximizing the value of the strategic action that results from the understanding of the key aspects of a company and the factors that affect an industry as a whole. On step eight, the SWOT analysis is recommended as the analytical and synthetizing framework to proceed with the evaluation of all the angles of a business, an opportunity or a new future. Applied correctly, an analysis of this nature can help organizations at different levels of performance to recognize and take advantage of relevant and sustainable market (internal-external) opportunities.

Chapter 3 (Steps 9, 10, 11, and 12), "From Vision to Action", suggest an approach based on the proposals of client development and design thinking, including the different aspects from the scientific method applied to entrepreneurship, demanding to move from assumptions to facts in the set of hypotheses behind any opportunity to create value, where observation and experimentation are the tools for the discovery and learning of the market segment, the value proposition, the channels, the mechanisms for generating income, the price or cost structure and, in general, the elements of the business model. Discovery and learning must always be validated by observation and translated into hard data. Agile entrepreneurship as design thinking refers in essence to the discovery of a business model validated empirically and in no other way, and its subsequent scaling, through a process of verification of the assumptions and assumptions of the project. There is a particular focus on sales and marketing, understanding that in a market economy the customer, and only he, decides the survival of the relevant designs.

Choosing and carrying out the strategic intentions is the axis Chapter 3; organized in steps, placing special emphasis on the innovation process, understood as the practical and systematic method of producing new concepts of products, services or systems that are marketed in the market at a steady pace in a certain industry. Lean Intelligent Innovation is the proposed model, the result of combining three widely recognized methodologies: Design Thinking, Quality Function Deployment and TRIZ. Unlike intuitive proposals, Lean Intelligent Innovation is a logically constructed method based on data, providing reliability from a repeatable and predictable structure.

Chapter 4 (Steps 14 and 15), deals with the importance of extended leadership in the organization. Teamwork, inclusion, the empowerment of collaborators, the correct alignment with the newly created or defined organizational structure and, above all, a clear base for measuring performance that allows individuals to be clear about the objectives and goals that have been defined for them; allowing them, of course, to move in the proposed corporate innovation culture. This is the chapter that best symbolizes the corporate entrepreneurship approach. By managing each strategic initiative as an enterprise in itself (rather than separated, coordinated with the rest of the initiatives as a real ecosystem), each team is given the opportunity to propose, design, implement, measure, correct and improve any action undertaken by their teams, which together make up the extended equipment.

In Chapter 5 (Step 16), we emphasis that, the concepts conceived must be validated by means of experiments, resorting to the development of "prototypes" of diverse nature according to the hypotheses put to the test, passing from idea to execution. The capacity for experimentation determines to a large extent the possibilities of revitalization and survival of companies. The proposal makes a leap from the obvious to establish that competition for relevant design and market competition in general should focus on the business formula and not on the company or the product, explaining much of the failed attempts introduced annually to the market.

In Chapter 6 (Step 17), six principles are offered related to the validation of innovation proposals based on the business formula, which are relevant for reducing the conditions of uncertainty and risk inherent in any transformation effort, insisting on the need to test key assumptions periodically, considering that "everything" can be tested through experimentation regardless of "failure", as long as it is carried out quickly, at low cost and constitutes a source of learning.

Chapter 7 (Step 18), related to agile operations, focuses on the ability to profitably reproduce innovation, capturing the value created by it. To a large extent, the proposal coincides with the concept of exponential organizations and the treatment of complementary assets, but unlike and in accordance with the appropriation regimes proposed by Teece, the possibilities will be a function of the nature of the technology, of the practical possibility of exercising the right over intellectual property and the relevance of complementary assets for innovation. It offers an interpretation of the company as a complex adaptive system of business formulas enabled by the business foundation and operational capabilities, the latter the sustenance of the creation, capture and multiplication of value.

Stragile offers a map for the renewal of companies in dynamic environments, particularly in areas of action where innovation is not a routine practice of specialized organizations in the best case, in continuous improvement.

However, the latter should not be dismissed: The change of rupture allows discovering new business formulas, but developing the market requires continuous improvement, efficiency to make resources productive and develop cost advantages. In the initial stage, creativity, experimentation and flexibility are required, with the sole objective of discovering and learning; in the development of the market, discipline, statistical control of the process and routines are necessary to give economic and growing results, closer to closed systems and less to open systems. In other words, change management is relevant when the business formula has expired and needs to be renewed; the administration of efficient continuity is relevant in the phases of growth and maturity. The company man must understand them like that and not in another way.

Finally, Chapter 8 (Step 19), of Conscious evolution, consists of committing oneself to a process of experimentation, iteration, learning and decision about the results at different moments in the development continuum. *There is no failure, only learning.* Through the iteration of search and selection, companies and their respective value chains evolve over time, with the conditions of the sector at a given moment, the seed of their condition at a later time. As a general rule of evolutionary systems, fortuitous cases and random causes exist, however, are the administrative decisions, the mechanism of internal selection of the organization, and the choice of the market, the mechanism of selection of the environment, which dictate the survival and configuration of strategic choices.

To conclude, we offer below some final recommendations for the implementation of the Stragile methodology in the established company.

1. Those responsible for the future of the company, generally the owners of it, have the responsibility to permanently evaluate the health and relevance of the company's business formulas. If innovation is about creating wealth, then early abandonment or late abandonment can be catastrophic. Having clarity of the potential for creation, capture and multiplication of value of each of the business formulas is an obligation for the company man.
2. Innovation must be integrated inescapably with the company's business strategy. It should not be experienced when the crisis has arisen, but constantly to identify concepts with the potential to become relevant designs for a market segment. Even when you do not have the "need" to replace or renew the business formulas in the present, the innovation

routines prepare the organization in the understanding that sooner or later the time comes to do it.

3. Ideally, to provoke transformation concepts alienated to the business foundation and the operative capacities to endow them with a new potential for wealth creation. When this is not possible, the development of new foundations and dynamic capacities will be necessary. However, the established company has a huge advantage over the startup, being precisely the possibility of focusing the foundation as the capabilities in indirect segments, extending the line of products and services, offering complementary products, introducing ancillary services, extending the chain of value of the industry or simply find new uses or applications for new attractive and growing segments.
4. Establish selection criteria for transformation and innovation proposals, particularly those that offer the possibility of producing profits or economic benefits resulting from business activities, for a market with sustained growth potential and potential for diversification, respecting the values not negotiable of the organization. Project evaluation techniques should specifically incorporate these criteria.
5. Finally, consider the possible contributions of technologies different from the company's own and traditional ones, particularly those close to its commercial possibility known as exponential technologies. According to Klaus Schwab, these can be grouped into three categories:
    a. Physical technologies, such as autonomous vehicles, 3D printers, new materials and advanced robotics.
    b. Digital, including Internet of Things (IoT), sensors, electronic data exchange, blockchain, artificial intelligence, the 'unlimited' capacity of storage, the processors of one or more terahertz and WiFi networks based on light infrared, and
    c. Biological, in particular molecular biology and synthetic biology.

The improvement of a company's capacity and performance, as well as its reduction in costs, is inevitable. Nonetheless, the potential lies not in the development of individual technologies, but in their convergence. To enable shared economy business models like Uber or Airbnb, until the aging of people, passing through fuel based on nano aluminum alloy batteries, adding only water. But there is another feature in these technologies: they are enabling new possibilities for any product or service. What's in them for the company?

Acknowledgment

To my parents, Jesús and Hermila.

To Magaly, Nicole, and Megan, my source of inspiration. – Jesús Enrique Portillo Pizaña

To my wife, Conory, and kids, Sergio, Luis Julian, and Cony, and all the people who have helped me to understand the innovation and entrepreneurship process. – Sergio Ortiz Valdés

To my beloved wife, Marisela Liselot, and sons, Luis Miguel and Luis Ariel. – Luis Miguel Beristain

A special recognition and thankfulness to our great team who made it possible: Jazmín Gallegos, Elías Chacón, Carlos Cortés, and Luis Felipe García, with a special distinction to Andrea Martínez and Tania Moreno, who were part of the beginning of this dream.

Chapter 1
Ignition

ABSTRACT

Every process of transformation—human, institutional, or community—begins with the stage of consciousness. Can we evolve without being fully aware of what we want to achieve? Of course! However, this implies going as a wind vane, waiting for the wind to indicate the direction to follow, with the respective risks and consequences of lack of control or defocusing. A second step involves making decisions, which implies risk, which generates fear, which ultimately paralyzes the decision-making processes; preparing ourselves consciously for this stage is highly relevant. The third stage, action, involves a lot of discipline; the problem with this is that comfort also paralyzes us as human beings. This is perhaps the biggest challenge to overcome. Transited the previous three, evolution involves two different types of challenges: 1) how to continue adding value to what we have achieved and 2) how to restart the cycle reinventing us again and again. This chapter explores these beginning steps.

STEP 1. CONSCIOUSNESS: IT ALL STARTS WITH AWARENESS

Introduction

I'll live this moment to its fullest, because if it's not me, then who? if it's not here, then where? if it's not now, then when? and if it's not this way, then how? (Jodorowsky, 2012)

DOI: 10.4018/978-1-5225-4023-6.ch001

Once one has crossed the consciousness line and contemplated new growth opportunities, once one has understood that any process of transformation starts with awareness and clarity of the path to follow (either personal, family, business, or community-related), then we have understood that the moment to make the right decisions has come.

A great friend of mine, Dr. Arturo Valenzuela, during the most turbulent years of insecurity in *Ciudad Juarez,* shared publicly a favorite phrase of his*:* "choose your thoughts." With this, he concocted a wake-up call for the community to re-consider the responsibility each one of us has concerning the decision we make.

On that moment, Arturo and I had a chance to talk about what was happening in our city and the hidden opportunity behind his phrase. I commented it appeared to me that there was more to it. For me, *"choose your thoughts"* just wasn't enough because the foundation of these thoughts could be distorted or misconceived. It was right there when I understood that every process of transformation of a human being, a family, an organization, or a community should start from a state of self-awareness.

The power of decision-making would be useless if the foundation on which this decision had been made was wrong. At the same time, being a group of citizens with a great deal of desire to overcome the situation would also be useless without a clear cognizance of the purpose (or goal) to reach.

Defining who I am and where I want to go is, and will always be, the starting point. This is when the second great element of this process of conscious evolution came into play: The Choice. At one point Jorge Bucay (2015), shared one of his greatest phrases:

Because no one can know for you, no one can believe for you, no one can seek for you, no one can do what you must do for yourself, and existence does not allow representatives. (Bucay, 2013)

Decide implies responsibility, ethics, commitment, and a social consciousness. A decision should be made under the basis of what's best for the majority; it implies a deep sense of maturity that can overcome the benefit and the welfare of the human being over any element of personal interest.

Then, a third element emerged. Consciousness and choice could never reach a tangible state by themselves if they weren't followed by the Action.

One can have a clear contemplation of the self, where one wants to go, the ability and clearness of thought to generate brilliant ideas or bright alternative ways to solve a problem, but the first step to transcend will always be the materialization of ideas through actions.

The disciplined, persistent, and—above everything—resilient execution of actions, will always lead toward reaching our goals.

The fourth element of this entrepreneurship cycle is Evolution. Pretending that every step we make will be successful is to ignore the fact that there is no perfect organization. There is enough literature to this day emphasizing on making failure a big part of success that we must understand that the outcome of our actions must not be defined by failure or success, but by success and learning.

The ability to learn quickly has become one of humanities' greatest virtues. The agility, equanimity, and resilience with which we face stumbles is now the principal tool to reach our goals in an effective and efficient way.

- To us the Cycle of: Consciousness (Discover), Choice (Decide), Action (Develop) and Evolution (Deliver) is the basis of this proposal of agile innovation strategy.

There is no coming to consciousness without pain. People will do anything, no matter how absurd, in order to avoid facing their own soul. One does not become enlightened by imagining figures of light, but by making the darkness conscious. (Jung, 1928)

Figure 1.
Source: Moreno, T. (2015), Entrepreneurs DNA.

During a different time of the abovementioned social crisis in Ciudad Juarez, and in the presence of the former Mexican President Felipe Calderon, Arturo Valenzuela suggested another idea:

Pain generated consciousness; consciousness generated action. (Valenzuela, 2014)

Just like Carl Jung, Arturo recognized to the former president, that in Juarez we have been deceiving each other for a long time.

We fell asleep over the success of the *maquiladora* industry; we thought that the successful economic cycle would go on for many more years. We were able to create a world-class industry that could export not only high quality goods, but also world-class talent. The big corporations were here in the region, and as expected around them came the restaurant, hotel, and tourism industry that inevitably adjuncts to a world-class economy. However, during that time, we stopped innovating. The flow of resources was such that the appearance of a new restaurant would not affect the economy of the already existing ones, which lead to think that opening more hotels, restaurants, or industrial environments to rent, would continue to produce results for a long time.

Meanwhile, inside the *maquiladora* industries, the innovation cycles were limited to quality-control processes. At one moment on a roundtable with some leaders of these great corporations I asked, "What are the innovation processes that prevail at your companies? They quickly responded, "Six Sigma," which made me ask: "…but those are continuous improvement processes, aren't they? And as far as I understand (as impossible as it may seem), Six Sigma has its limits," once again one of them answered, "In fact, the complication is that we have reached a level of excellence so high that we have reached those limits." Then we understood that also inside those big corporations, innovation had stopped adding value to our economy. To make matters worse, when initiatives of innovation or a patent emerged, the registry of that patent was made at the country of origin of the company. This meant that the capitalization of the regional human talent was also slipping away. Also, as if that was not enough, we were unable to find a world-class supply development plan at the moment to substitute the flow of imports coming daily into our economy.

When we had to face the pain of the economic and security crisis was the moment of realization that we had fallen in a sin by omission. We got used to the constant flow of product and services (both lawful and illicit) and the stability level was such that we never reach to understood that the social stability was at a boiling point.

To the group of people (myself included) that created the concept "Stragile", this was a defining moment, (understanding defining moment, as every moment in our lives that forces us to make a choice and that define us as persons).

Another important moment of reflection came when, in the face of the city's security crisis, many chose to abandon the city, others barricaded at their homes, businesses closed, and fear was allowed to prevail in all of us.

One of my working partners asked during these times, "*Aren't you afraid of being working on these security-oriented matters?*" To which I answered, "*I am not afraid, I am terrified*", but right then Jodorowsky's phrase came to my mind, so I closed my comment by saying, "*...but if it's not us, then who?*"

It was at that moment when we fully realized that any process of human, organizational, or community transformation, begins with a level of awareness that forces us to make choices, obligates us to act, and offers an opportunity to evolve.

We are used to let someone else make the decisions for us, we wait– comfortably— for someone else to tell us what to do; this way we have the perfect and irrefutable excuse (in case the decision was wrong) to say "*It wasn't me, I was just following orders.*"

The idea behind Arturo's phrase was exactly that. You are responsible to make every one of the choices that have to do with your life. No one else will or should do them for you. While it is true that living implies risks and sacrifices, you can't have it all, and generally uncertainty comes with fear, but it is in these crossroads when the responsibility to choose implies to let go (or leave behind) at least momentarily, the untraveled path. To make the right choice is hard, but it will be the one that allows for fertile soil to grow.

The Recovery

When the storm finally vanished thanks to the community's efforts to move on (as it usually happens in similar situations) the next question was: *What do we have to do now in order to build solid bases for the future economic growth and social wellness, all under the concept of human dignity?*

At that time, a small group of (3) businessmen extended us an invitation to participate in the development of a technology park for the city. To them, the solution was in apportioning infrastructure crucial for the development of innovation.

I believed that the creation of a technology park wasn't necessarily the answer we were looking for. Rather than the creation of a *white elephant*, Ciudad Juarez needed urgently to detonate its associative ability to generate a real innovation and entrepreneurship ecosystem, based in the social wellbeing. To which arose the alternate proposal of developing a Research and Integration Center of Advanced Technologies that could work as the *"coordinating brain"* of the city's innovation strategy; allowing us to capitalize each existing and soon-to-be-developed technology investments in the city without resources or labs being repeated and without them professionally overlapping. On the other hand, the objective was to allow that the competences that derive in the city were coordinated in a way that they could complement each other.

The strategic approach we developed was centered on the model known as "*The Triple. Helix,*" which proposes to coordinate all government, business, and academic efforts. At the birth of the strategy, this initiative was tagged as foolish. For many of the participants, the mere act of thinking about it formulated negative answers everywhere. Against such resistance, the question was: *How can we make this happen?*

A small group of researchers from local universities helped us find ways to shape and make the proposal work. Of course, the main goal of this project was to create the public policy to increase the investment on science and

Figure 2.
Source: Portillo, E. (2016), Advanced Technologies Research and Integration Center (CIITA).

Figure 3.
Source: Portillo, E. (2016), Triple Helix Integration Model

technology education, aligned with the recognized region's manufacture industry needs.

This way, the ability to associate we could trigger, should generate public policy to increase the investment of scientific-technological infrastructure needed to support the innovation processes and consolidate us as local suppliers.

During the process, we started to find models related to innovation under the concepts of learning theory proposed by David Kolb, Roger Greenaway, Chris Argyris, and Donald Schon; which proposed four steps, as shown in Figure 4.

We also learned that, in order to make proposals this big, we were to become inclusive of other three concepts: The technological feasibility of the proposal, the viability of the business model, but most importantly, the creativity taking into account the needs of the market.

To accompany the process of creation of the Advanced Technologies Research and Integration Center (*Centro Investigación e Integración de Tecnologías Avanzadas, CIITA*), we used what we learned at the *Tecnológico*

Figure 4.
Source: Moreno, T. (2017) Based on: Batista, Ed (compiled by Andrea Corney), (2007).

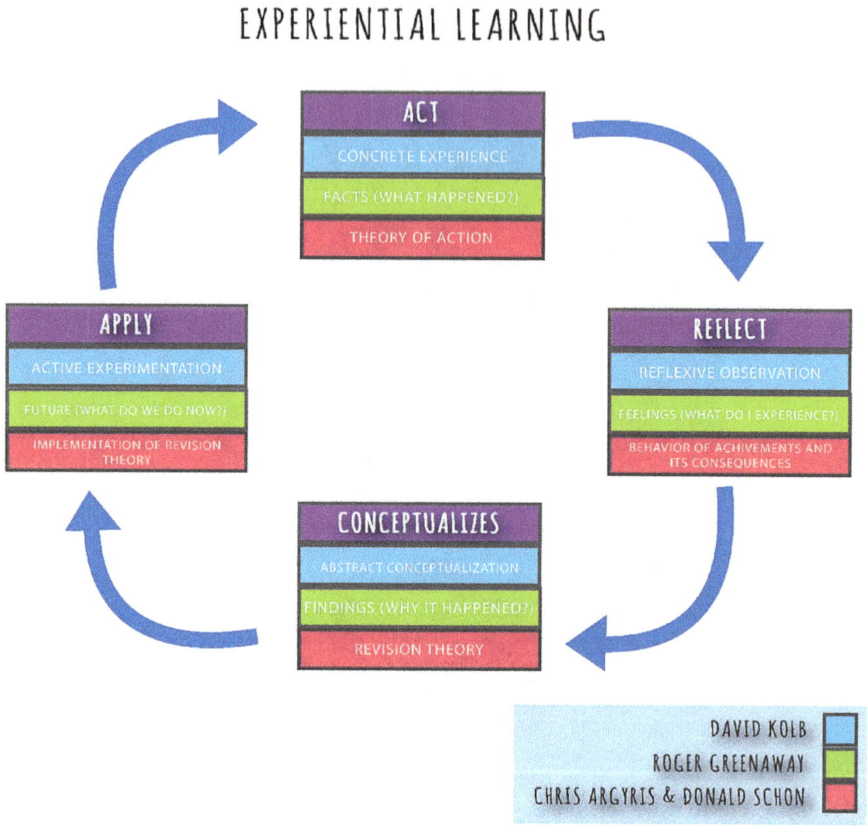

de Monterrey; Monterrey Institute of Technology and Higher Education regarding different innovation models that prevail at the world-class economies.

Theories and Models like the Business Model Canvas, the Lean Canvas, Steve Blank's proposals, and all that have emerged recently, (including the traditional marketing strategy), gave rise to the foundation of the 4 stages of the Stragile model:

- Consciousness (Discover),
- Choice (Decide),
- Action (Develop) *and*
- Evolution (Deliver).

Figure 5 shows the way some of the most relevant proposals are categorized.

Figure 5.
Source: Portillo, E. (2016), Stragile base theories.

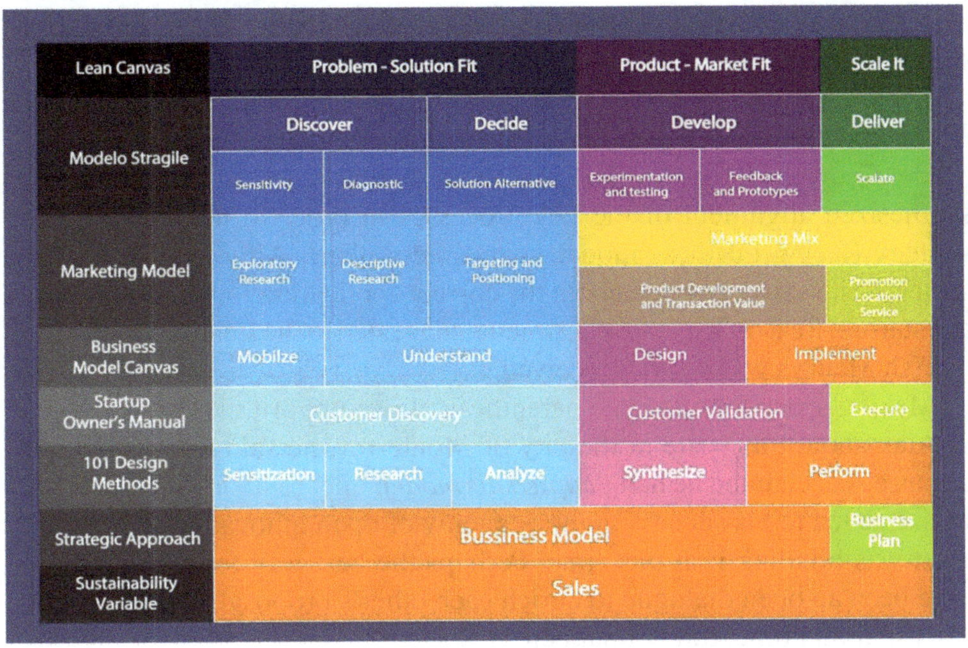

When we connected the traditional marketing model with the entrepreneurship models, we realized that there were certain coincidences that allowed us to summarize all types of activities to three basic stages within the innovation and renewal cycles of organizations. Stages whose author Ash Mauraya defines as:

1. Problem-solution adjustment.
2. Business-product adjustment
3. Scaling.

We consider that any process of strategic development of a product or service goes through these three stages. The importance of better understanding the problem or opportunity presented to us is key. Like the saying goes, "a well-defined problem is an *almost* solved problem." Unfortunately, this is not enough. It is necessary to immerse into a process of corroboration of our value propositions so that we do not make strong investments before the first sale has been made. It is precisely *the first sale* the main objective to fulfill before taking it to the next step.

In the sales process, the market will instruct us on the adjustments we must make to our products or services. It is then and *only* then, when we will be ready to develop the rest of the strategies needed to scale our business strategy. This is how the three phases are fulfilled.

Up to this point, the focus has been on models related to the business incubation approach—with the exception of Alex Osterwalder's Business Model Canvas (2010) whose success was due more to the simplicity of its synthesized alignment of the key processes in order to detonate an agile strategy (facing business model vs. business plan). Although, in its early stages, it showed confusion about the correct application for the initial stages of a startup. Reason why the adjustments suggested in the Lean Canvas (2012) by Ash Maurya was so well received.

Making space for anecdote parenthesis: In an opportunity we had to talk to Alex about this issue in the city of Monterrey, he said he was annoyed because Ash claimed he had *"copied his model,"* not realizing that, precisely what had led to his success in the economy of open innovation, was exactly what was blinding it at that time. Perhaps the best way to respond should have been an open dialogue to enrich each other's proposal, a situation that, at least at that time, did not happen.

We argued that the Business Model Canvas (BMC) was, perhaps, the tool whose approach had more weight in advanced companies. And it is not that there was something new for the stages of acceleration and consolidation of a business. On the contrary, for these stages, there have been many business management theories developed during the years. You could name whichever you want, but the BMC definitely fit into the business plan stage!

Nevertheless, it should be noted that, within all competitiveness, excellence, quality, and reengineering theories, there is a recent one. The Kotter Model (1996), which fits rather positively with the nature of startup development, but as seen under the spinoffs approach.

Kotter's Model of Organizational Change presents a methodology that helps in the strategic structuring for the launch of initiatives in corporate innovation, business reengineering, repositioning of brands, or more appropriately, the term Stragile which we want to use to explain business strategies that generate new schemes of work in the face of change in an organization. The model is included in Figure 6, trying to find the process of alignment (methodologically speaking), with all the abovementioned, but for advanced stages of the life cycle of an organization.

As we can see, both the Stragile model and Kotter's model for organizational change are strategically aligned in the four proposed stages of Awareness, Choice, Action, and Evolution.

Given the ocean of information to which we have access in our daily lives, one of the greatest skills to develop is the capacity for synthesis. In this sense, a Canadian passionate about entrepreneurship is Alex Bruton. Alex managed to synthesize brilliantly and graphically every single one of the models that existed in the market at that time, allowing to satisfy a need that most of the people that work in this field required: a linear scheme.

While it is true that the current of entrepreneurship theories suggests that this is not a linear process, it is also true that detaching ourselves from a learning process that has shaped us for so many years has not been easy. It is clear that understanding the problem does not follow a linear process, but our mental scheme still requires us to find one.

Figure 6.
Source: Portillo, E. (2016), Stragile Corporate Innovation Process.

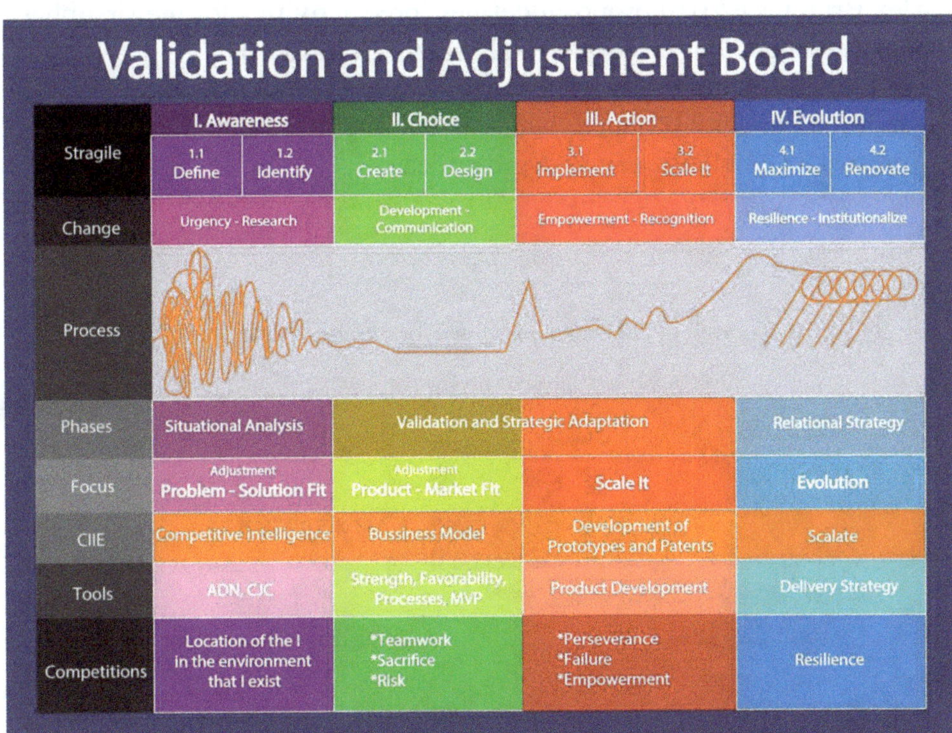

Going back to the graph that already includes the model of Kotter, we can see (highlighted in blue) how the stage of consciousness is linked to the need for the collection of information that supports and allows the correct decision-making; what we call competitive intelligence.

The action stage is then, associated to a concept of business model development. A process linked to the validation and reinforcement of the value proposition, as well as the exercise of a profitable financial model.

At the *action stage*, we understood that moving from business ideas to the development of prototypes and patents was of the utmost importance. We also learned that is disruptive and tangible products and services the ones that will make the difference in the market. If there is nothing of value in the offer, it is better to return and start again before betting all the capital on proposals that no one, not even our closest family and friends, would buy. At the last stage, scaling, it is necessary to understand and separate at least two phases: growth and maturity.

A disruptive product or service, usually in its beginning stages of introduction to the market, starts alone and without any competition, but once the market reacts, the blue ocean becomes a red one and the complexity of differentiation becomes more serious.

Alex Bruton's (2016) proposal defines three activities for the first phase of an entrepreneurship project at the Problem-Solution fit:

1. It all begins with a hunch, which is nothing more than the awakening of, what we believe, can be a great idea.

Figure 7.
Source: Moreno, T. & Cortes, C. (2016), Adapted from: Alex Bruton (2016).

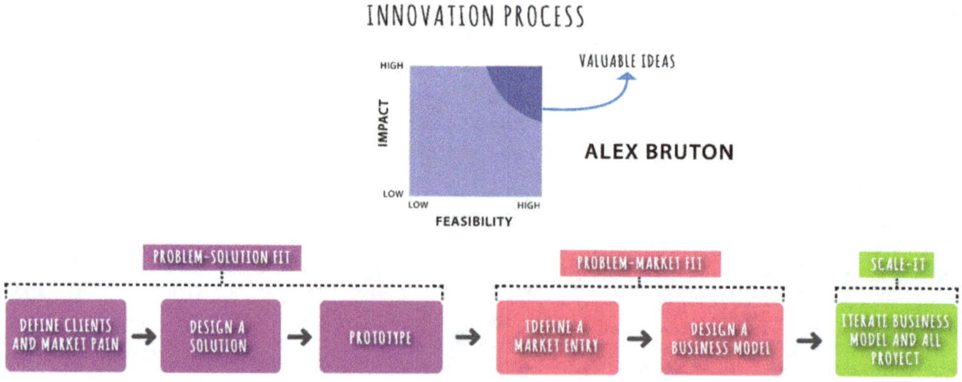

2. Then, he proposes the idea of the napkin, which explains that we will usually develop our mental maps on a sheet of paper, a paper napkin or whatever we have on hand to draw our idea.
3. Once we develop our mental model, the third step in the problem-solution adjustment stage is *the sketch*, a more evolved and better grounded version of our napkin drawing.

For the Product-Market fit stage, Alex suggests the application and combination of two approaches, understanding that at this stage the priority is to impose a proposal that can quickly adapt to the needs expressed by the market. The focus for this stage would be centered on startups or spinoffs.

Bruton gives order to the use of the Canvas models for the first time; suggesting the use of Ash Mauraya's Lean Canvas model—which focuses much more clearly on the need to have a well-defined problem, allowing clarity in the promotion of a creative solution alternative.

Once progress has been made in the process, the next step proposed is the need to use the Business Model Canvas to focus on the unique product strategy towards an efficient business model adapting to the needs of the target market. Just when the product or service has been validated by the market repeatedly, its level of acceptance will guarantee the sale.

Finally, Bruton puts the business plan in motion. The comparison between the business plan and the Business Model Canvas, sometimes got confused, thinking that they were the same concepts or that there was no differentiation between them. It was also believed that one was being displaced by the other. Although our perception of both concepts was wrong, the business plan has not and will not stop to be required. However, its usefulness is understood more as a tool of approach and follow-up towards execution.

By developing the alignment exercise for each one of the models that were in the market and coming across the Alex Bruton proposal, we understood that: although it is true that any company is linked to linear processes for chronological reasons, it is also true that all personal, family, business, or community projects live non-linear cycles of progress and setbacks.

We understood that these three stages of: Problem-solution fit, Product-market fit, and Escalation, were aligned with the cycles observed in other methodological proposals, and that each completed cycle meant a process of evolution, since people, families, companies, or communities would necessarily have to be wondering what would come next. This is where the conscious evolutionary process comes into play.

Along with every one of these theories and studied models, we found another reference that comes to complement the above. In the book *The Art of Getting Started* (2004) by Guy Kawasaki, suggests different phases in the process of launching a company, with different challenges for each:

1. **The Cause:**
   a. The meaning or purpose.
   b. The mantra instead of a complicated mission to remember.
   c. The suitable launch instead of an elaborated business plan.
   d. A well-defined business model with an emphasis on financial profitability.
   e. Milestones, bases and tasks, which is nothing more than the plan of action.
2. **The Joint:**
   a. The art of positioning the value proposition.
   b. The art of presentation or pitch.
   c. The art of the written business plan, now as a sure guide to execution.
3. **The Activation:**
   a. Bootstrapping, starting from scratch with the resources you have at hand.
   b. First to sell and then improve, before investing in a large corporate structure, the first challenge must be to sell and capitalize on growth. From our point of view the most powerful concept of all!
   c. The art of recruitment involves attracting human talent that matches the organization's DNA and that is competent in the skills required.
   d. The art of seeking capital, implies laying the foundations on which to consolidate the company's strategy of escalation.
4. **The Growth:**
   a. The art of association, an association strategy will always be advantageous to consolidate the process of expansion of the company.
   b. The art of creating a brand implies taking care in detail of the famous 4 P's of marketing—to which we add the "s" of service, for the importance that it represents to develop strategies of retention and loyalty.

The Art of Getting Started coincides with the previous proposal models alignment; alignment centered in the stages of Problem-Solution fit, Product-Market fit, and Escalation, with the difference that Kawasaki defines his model in four stages and that, by its nature, relate to the stages that have been defined in each one of the models previously presented:

1. The Cause = Problem-solution fit.
2. The Joint = Internal adjustment of Product-market fit.
3. The Activation = External adjustment of Product-market fit; validation.
4. The Growth = Escalation.

All the coincidences in previous approaches allowed us to conceptualize (again) our model under the four-step approach: Consciousness, Choice, Action, and Evolution which of course required to be broken down with much greater clarity.

During this breakdown, we understood that the tools could be used for a company in its early stages, as well as for companies in a stage of maturity. We understood and developed the Stragile process for the neophyte companies and, at the same time, we found the model of Kotter worked for the leadership facing change, for already-consolidated companies.

During this process of "connecting the dots" between the different models of entrepreneurship, the marketing literature and the classics of business literature, we found yet another powerful connection with one of the books that revolutionized the current approach to corporate culture. *The Fifth Discipline* by Peter Senge (1990).

In this masterpiece, Senge proposes four central disciplines: Personal Domain, Mental Models, Shared Vision and Team Learning, which in the end are coordinated by The Fifth discipline: Systemic Thinking.

1. **Personal Domain:** It implies transcending certain traditional labor schemes. Going beyond the mastery of competences, skills, and values required by our profession. It also suggests a deep maturity and a collective consciousness; which makes sense for the application of soft skills with a creative and non-reactive approach to human meaning.
2. **Mental Models:** An intelligent, active-learning organization must understand and respect the fact that each one of its members has different paradigms. The truth of one member will surely be different from the truth seen by another. Thus, we have a general understanding that it is not an absolute truth that dominates the contexts in which we live; there are three: your truth, my truth, and the absolute truth. Communication and respect in this case gives rise to the creativity of the human being. One of the main reasons why we have questioned the traditional processes

Figure 8.
Source: Moreno, T. (2017), Adapted from: Senge, P. (1990).

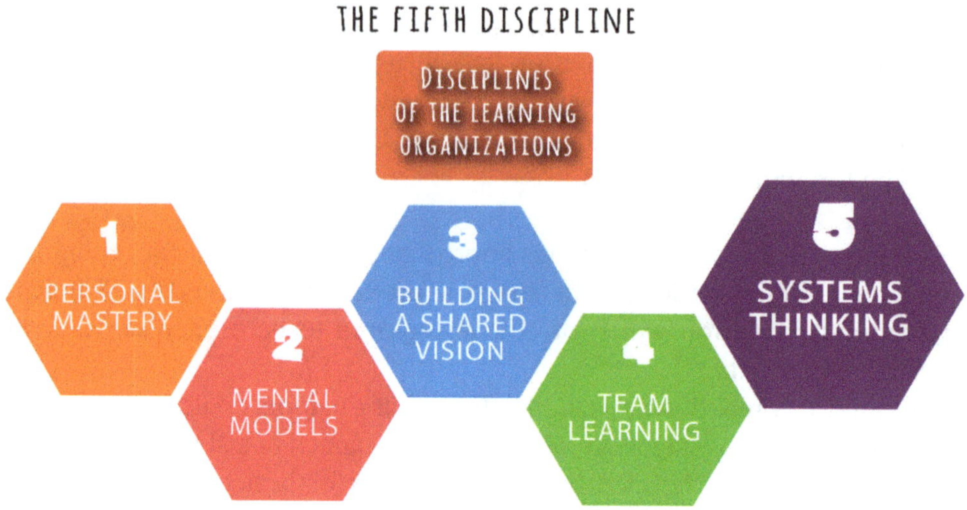

of strategic planning is this point in particular. We realized that, in these processes, most of the creative ideas were adrift either because at the developmental phase (those who did not understand the proposal did not embrace disruptive ideas), or because who led the organization ended up saying, "do not worry, although the idea has not been approved, is something that we must do," which conditioned innovation to the criteria of the leader, instead of the *intelligent* organization.

3. **Shared Vision:** Refers to the importance of having a common vision to efficiently achieve the goals and objectives of an organization. We have firmly held that every process of transformation begins with a state of consciousness that defines who I am and where I want to go. We understood that by sharing publicly our goals and objectives, as we seek acceptance and agreement with others and finding them, the bonds of collaboration are strengthened and the passion increases. This is one of the main sources of motivation for every human being.
4. **Team Learning:** In his book, Senge shares the vision of a key player of one of the most impactful teams in the history of professional basketball: Bill Russell of the Boston Celtics. In doing so, Senge finds the perfect analogy to explain the basis of true teamwork. The coincidence of visions and the passion to transcend are not enough in an intelligent organization,

the organization requires complementation and a coordination of skills and abilities to transcend the achievement of shared goals.

5. **Systemic Thinking:** The harmony of an organization will depend on the acceptance that the sum of the parts is more important than the individual interests. The current theory of business management has questioned in different ways (and at different times) the formation of *silos* by the power struggle. The motivation theories explained below will help understand human being's fears in matters of personal interaction.

The Learning

To close this first part of chapter one, we want to introduce for the first time the concept of the Stragile model (Figure 9). In it we can see different concepts from diverse authors and different approaches, all of them merged in such a way that they coincide and help us explain the four stages of the Stragile foundation: consciousness, choice, action, and evolution.

We are proposing in this model, a circular process rather than a linear one, with the idea of the evolutionary cycle. Therefore, the model involves the following steps:

Figure 9.
Source: Portillo, E. (2016), The Stragile Conscious Corporate Innovation Model.

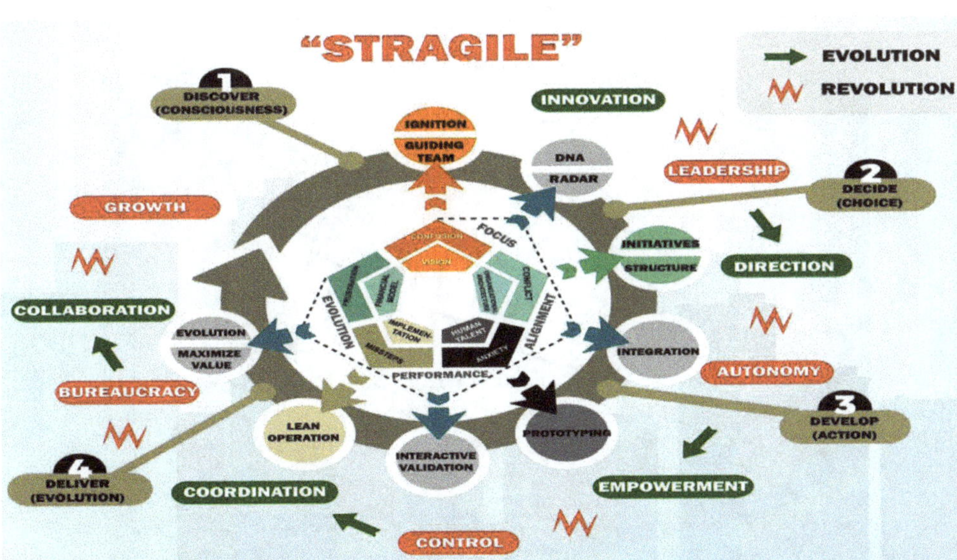

1. Ignition
2. *DNA* Definition
3. Creation of strategic initiatives
4. Inclusion of human talent process
5. Human talent integration process
6. Interactive and collaborative validation
7. Focus on sales operations
8. Value maximization
9. Evolution

The model includes one of the key concepts to understanding the evolutionary process of organizations proposed by Larry Greiner in 1972 *Evolution and Revolution as Organizations Grow*, and reintroduced by Harvard Business Review in 1998 (Hbrorg, 1998) (Figure 10).

The Greiner Model, which will be explained in depth shortly, proposes a fundamental idea of the process of growth in organizations, a concept that we have adopted to understand any evolutionary process of a human being, a family structure, an organization, or a community. It suggests the idea

Figure 10.
Source: Moreno, T. (2017), Adapted from: Larry E. Greiner (1998).

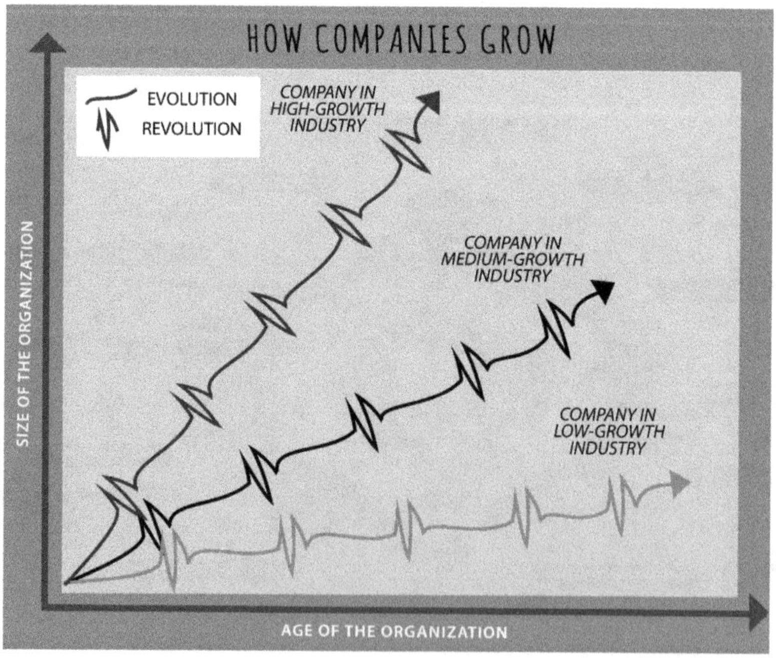

that behind every crisis or revolution involves a growth or evolution. These stages of crisis and growth have been represented along the Stragile model, suggesting the different moments in the stages of growth of an organization.

An additional element that has been included in the Stragile model, is one that solves confusion with vision (what we call focus); resolves the conflict with organizational architecture aligned with the inclusion of human talent, what we call alignment; replace the false steps with the implementation of processes, what we call execution; and finally, proposes the maximization of value and the appropriate management of a financial model to give way or to "*set the stage*" to properly scale the business model, what we call evolution.

It is convenient to emphasize, although it seems obvious, that this model is a generic model and that necessarily requires adaptations depending on the case being evaluated at the time of its application. Comparing a company in the acceleration stage, against a company at the incubation stage, would be aberrant.

To clarify this last idea, we show Figures 11 and 12 that help explain the application of the Stragile model in two different cases. An example for a company at the incubation stage, and a company at the acceleration stage. The model generically presents different steps. It must be understood that there are different tools, models, and approaches that can be used at the discretion and convenience of who is managing the process of strategic adaptation.

There is a great variety of books that contain different tools and models. One in particular, *101 Methods of Design* (2012) by Vijay Kumar, groups and synthesizes different tools for its application at different stages and with different objectives, which we encourage (if you wish) to explore different tools to those suggested by the authors in this editorial work. It is necessary to clarify that it is not the purpose of this book to enter the discussion of which tools are best for each case, for it would be an endless debate given that each company seeks different objectives under different contexts and criteria. The same happens with the criteria of who is applying the tools. Those who feel more comfortable understanding and using a given tool, will surely defend the basis of their application, without giving the opportunity to explore other options. The graphics used to explain the model show we have adapted the names of each stage depending on whether it is for an incubating company or an accelerating company.

Figure 11.
Source: Portillo, E. (2013), Stragile at startups and Accelerators.

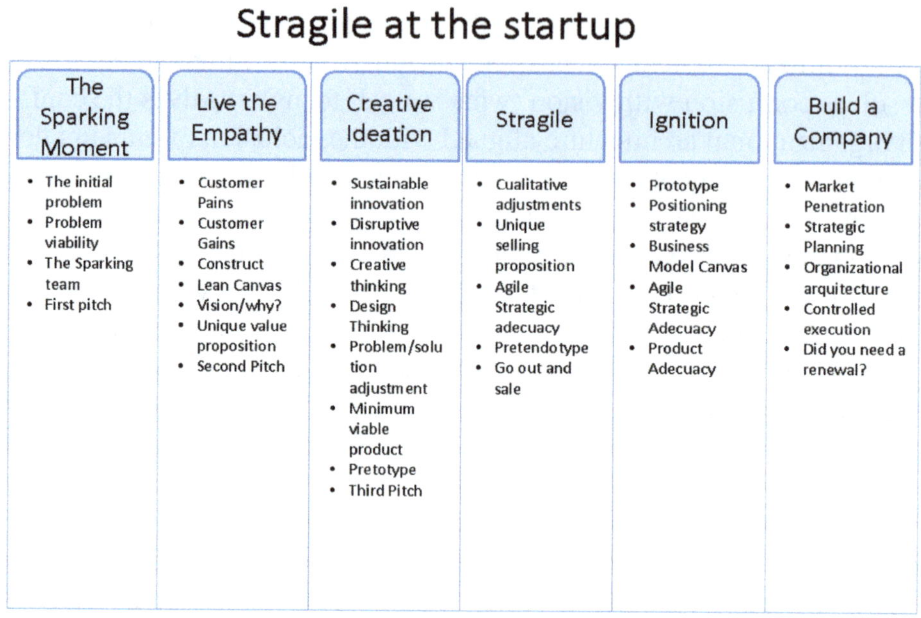

Figure 12.
Source: Portillo, E. (2013), Stragile at startups and Accelerators.

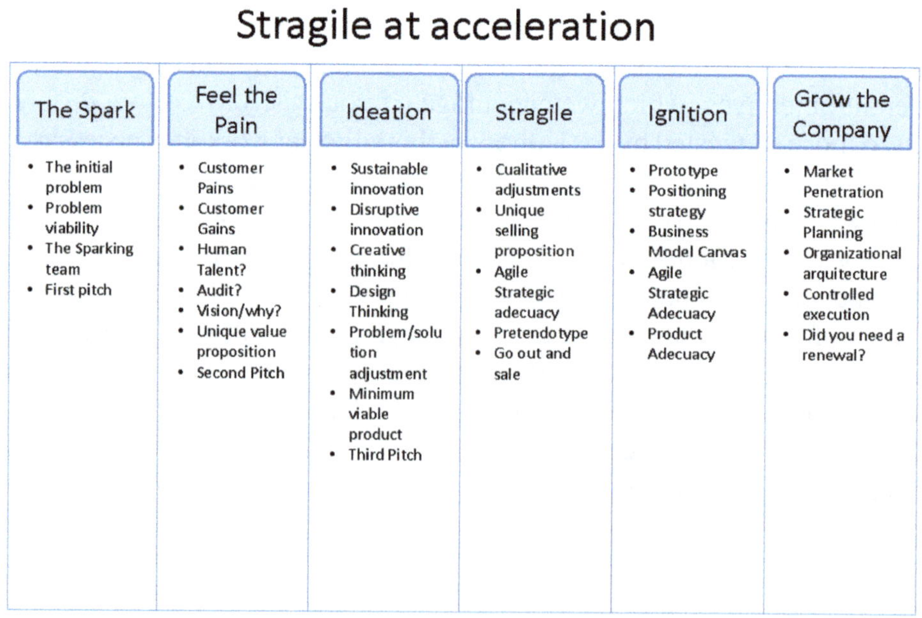

STEP 2. FIND THE SOURCE OF YOUR INSPIRATION: MOTIVATION

We who lived in concentration camps can remember the men who walked through the huts comforting others, giving away their last piece of bread. They may have been few in number, but they offer sufficient proof that everything can be taken from a man, but one thing: the last of the human freedoms—to choose one's attitude in any given set of circumstances; to choose one's own way. (Frankl, 1946)

Explaining human behavior is fairly complicated, but describing the fundamental factors that motivate human behavior is much more difficult to explain. The key element that triggers any initiative, project, or transformation process is human motivation.

Finding the reasons that influences or that activates the passion of the human being is the key to any entrepreneurial process. Understanding how a human being can be deprived of his freedom in such precarious conditions, confronting death every day, and still able to maintain hope, clarity of thought, and above all, the correct attitude to succeed is fundamental to define the processes of motivation that the human being follow before the challenges and opportunities that life presents to us.

On one ocassion, some students asked me the difference between success and happiness. A student wanted to define his life purpose more precisely, but was faced with this dilemma. Is success the same as happiness?

If we look in the dictionary, surely we can find the definition and solve the existential crisis of this student. However, aware that this was not what the student was looking for and that, in any case, he required more support from his mentors to visualize himself in his personal context, we tried to find an answer that would give meaning to his life.

"We define success as having everything you want, but happiness is enjoying everything you have." In the end, we reflect on a different approach:

At this moment, there are currents of psychological and educational research, which propose that the human being should aim to achieve his/her happiness. In the strict sense of the definition, we could wrongfully assume that achieving happiness would mean enjoying everything you have—even if you have little to enjoy. However, we understood that true happiness for the human being is reached through transcendence. This is what the different psychological currents suggest to us at this point. Later we will mention some of the objectives pursued by these trends. To immerse ourselves in the topic of motivation it is imperative we start with the concept that yields the greatest motivation in the human conciousness: fear.

Several papers that have tried to establish the implications that fear has on our decision-making processes. However, one of those proposals interested us the most, not only for its simplicity, but for its great human sense. It is the one that Gerald Jampolsky suggests in his book *Love is Letting Go of Fear* (2003).

In his work, Jampolsky raises a rather interesting paradox. Generally speaking, humans act by either love or fear. It is the factors and circumstances that we have experienced that build paradigms, opportunities, and/or barriers in our minds. Our mental schemes based on what we have lived and every piece of information that we accumulate during our life are those that condition our perspective of reality. Before each stage, our brain consciously or unconsciously evaluates the situation to which it is been exposed, evaluates the risks or benefits existing, and if in the given situation this calculation is notde possible to make, then instinct prevails before reason. If the result of the evaluation suggests an opportunity, in most cases the entrepreneurial mindset begins to explore the possibility of making things happen.

Precisely for this reason, Jampolsky suggests that each of us is the author of his own destiny. Our mental maps only serve to prevent us from seeing, hearing, and experiencing love in ourselves and in others. This formation of instinctive and self-imposed barriers keeps us stuck in a retrograde belief system that we resort to daily, even when it does not provide the answers we are seeking.

Fear always distorts our perception and confuses us about what is really happening. Before a situation is detected as risk by our brain, fear can paralyze or activate the defense mechanisms necessary to protect us.

On the other hand, Jampolsky defines love as the total absence of fear. Love does not question anything. It is a natural state of extension and expansion, not one of comparison and measure. Love implies openness, understanding, empathy, and above all, respect. Love then, is the element that adds value to our existence, the desire to serve/help others in order to transcend is manifested through love. On the contrary, the absence of love is manifested in a defense mechanisms that does not add value to our lives.

Unfortunately, a poor communication process that gives rise to human fear can lead to various problems such as lack of confidence, disrespect, gossip, lack of commitment, power struggles, lack of integration, demotivation, and bad attitudes; crippling the team from establishing loyalty links.

One of the books that exposes clearly the problematic of the human paradigms and their effects in the productivity, is *Leadership and Self-Deception* (2010), written by the Arbinger Institute. In it, the main characters give life to the problem of human communication from the personal perspective, but from a professional angle (The arbinger institute, 2010).

In the book, authors treat the issue of personal choice as the key to assuming (or not) the right positions in the different scenarios in which we interact as human beings. They explain that, ultimately, each and every one of us is responsible for choosing the attitude and position we assume at each moment. If confronted with a problem, we can react violently or we can choose to ignore the accusation and inquire with equanimity and patience the situation before us.

The hypothesis of self-deception establishes that, in an incorrect position where the individual self-traps, we see ourselves as the victim; the one that gives everything, the righteous good partner, good manager, ideal workers. Just as we tend to deceive ourselves by justifying our actions and increasing our value in the human relationship, we begin to demean others in such a way that we see them as lazy, insensitive, ungrateful, inconsiderate, inept, and foolish.

When we betray ourselves, we tend to exaggerate our own virtues and, at the same time, the defects of others. What is worse, tend to justify against any argument that is pointed out and transfer the blame to co-workers.

Together, both books help us understand the negative impact that our fears and paradigms have on communication processes and, consequently, on the productivity of our projects.

Motivation Theories

As we have explained in advance, this does not pretend to be a textbook, on the contrary, it tries to establish its thesis by joining several elements related to the General theory of business strategy.

One of the key elements to fulfill this objective is precisely to understand the basics of the different motivation theories that ignite human behavior. In Table 1 we share some of the developers along with their key factors.

Of these theories, there are two in which we would like to focus. Most of them propose very interesting motivational factors—perhaps we should expand on Jung's approach as well (based on the theory of human beings visualizing future goals in order to reach it). It is also true that for the purpose

Table 1.

Author	Key Factor
Freud	Me, the ego, the super ego, sexual motivations
Jung	Future goals
Horney	Anxiety
McDougall	Instincts
Skinner	External environment
Hull	Drivers
Maslow	Human needs
McClelland	Achievement motivation, power motivation, affiliation motivation.

Source: Portillo, E. (2017), Summary of Psychological Theories.

of this book, the following are better aligned with the Stragile methodology proposed by the authors.

It is necessary to emphasize that Maslow's theory mainly helps to understand the needs of the human being that set off a behavior. Usually, Maslow's pyramid has been associated with the buying behavior of the external customer, for which it has been used much more frequently in the development of marketing strategies. However, as Figure 13 shows, the needs defined by Maslow applies in different circumstances of life. Maslow argued that one could not advance to the next level if the previous need was not satisfied.

Figure 13.
Source: Moreno, T. (2017), Adapted from: Maslow's Hierarchy of needs.

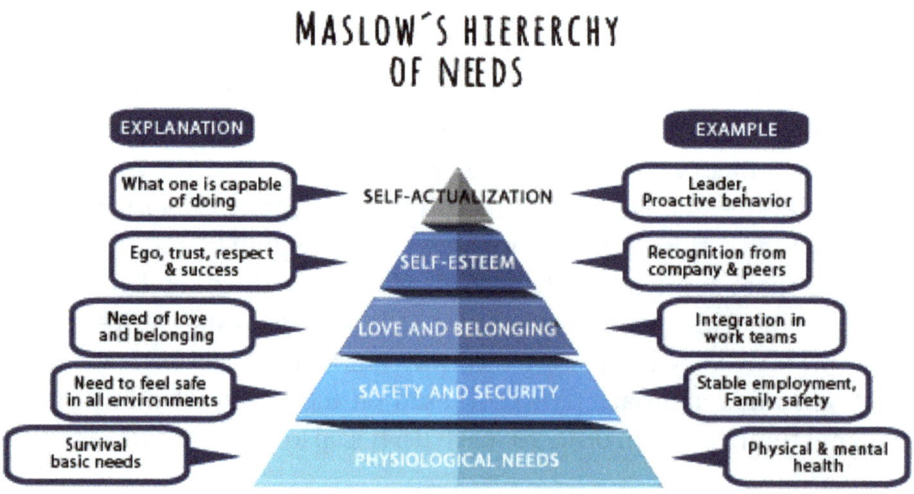

In Maslow's approach, we found that the first motivation of the human being is to satisfy his physiological needs; once these have been met the next step will be towards the satisfaction of their security needs; when the person has met needs such as stable employment or family security, the next level will be seeking social acceptance. Variables such as ego, trust, respect and success (among others) are in the category of esteem, there when the human being has already found certain levels of success and stability. The highest level of satisfaction is found in the self-esteem when the person seeks to transcend from the impact and recognition of his works. His legacy then becomes important; the person no longer experiences the fear of going down in the pyramid, so his focus is centered on the legacy of his work.

Comprehending Maslow's pyramid is relevant to understand what motivates each human being according to the stage or phase of life he/she is experiencing.

To further support the claim, we decided to use McClelland's theory of needs (2008), for we believe it represents more closely the motivational factors within an organization. As we can see, there are three needs that McClelland suggests every individual possesses:

1. **Need for Achievement:** It is the one that refers to the effort of the human being to excel, to achieve success from the task achieved, the pursuit of perfection in a job well done becomes the main motivation. At this stage the individualistic approach prevails over the approach of affiliation with other people.
2. **Need for Power:** It refers to the human need to exercise some kind of control in others. In this phase, the person enjoys being considered important. He/she seeks prestige, but also the imposition of his ideas. The predominant profile is that of the politician; one who is right at all costs, which gives him the power to exercise and decide.
3. **Need for Affiliation:** It is the need to relate to other people; to be part of a group and/or maintain contact with others and help other people. "Loneliness bothers me because there is no one to interact with or help."

These psychological theories referring to human motivation and behavior serve this book in the sole purpose of emphasizing the fact that human motivation is the true trigger of an attitude towards the manifestation of a particular behavior. What is relevant then, is that each individual defines his/her passion and has the ability to share it with other people in order to generate a shared vision, achieving common goals with dedication and commitment.

Crisis and Growth Processes

Human beings are not born once and for all on the day their mothers give birth to them, but ... life obliges them over and over again to give birth to themselves.(Garcia Marquez, 2007)

We have repeatedly witnessed how, in their childhood, our children suffer from a common cold. Treatment requires rest, a healthy diet, care, and medicine. We have also observed how during those periods of time, our children grew a few more inches.

In his article *Evolution and Revolution in the Growth of Organizations*, Larry E. Greiner (1998), introduced these two concepts and argued that during the different stages found in the life cycle of organizations, there were different moments with opportunities for growth.

Crisis, as we all know, can mean growth opportunities or facing the risk of disappearing from the market.

The innovation or ability of a company to ignite creative alternatives of solution will define a state of either grow or perish. In his work, Greiner recognized five phases of growth, as well as five stages of crisis.

From a lineal perspective, the process of innovation begins with a great idea; a creative idea that adds value to face the growing challenges of an organization.

Great ideas confront their first crisis in the face of resistance to change. Usually when someone has the ability to identify different solutions to a given problem, people who lack this ability tend to question them without understanding that this is an error of appreciation—not of bad faith, since if it were, then it would not be categorized as error, but as fraudulent or malicious. Those who cannot see the whole picture will question the reality and feasibility of those ideas.; the negative response received at this stage are our daily bread. The leadership crisis is then the biggest challenge to succeed in the process of growth of a company.

Leadership capacity, on the other hand, will allow us to enter the next phase of growth. Once the first crisis has been surpassed, a vision shared with strategic approach will give direction to the organization.

The next challenge then will be a crisis of autonomy, where the main actors, mainly subconsciously, will experiment power struggles to demonstrate their abilities and competencies; demanding the freedom to exercise their ideas and respect while believing themselves to be holders of a single, absolute, and permanent truth.

Figure 14.
Source: Moreno, T. (2017), Adapted from: Greiner, Larry E. (1998). The Five Phases of Growth. Evolution and Revolution as Organizations Grow.

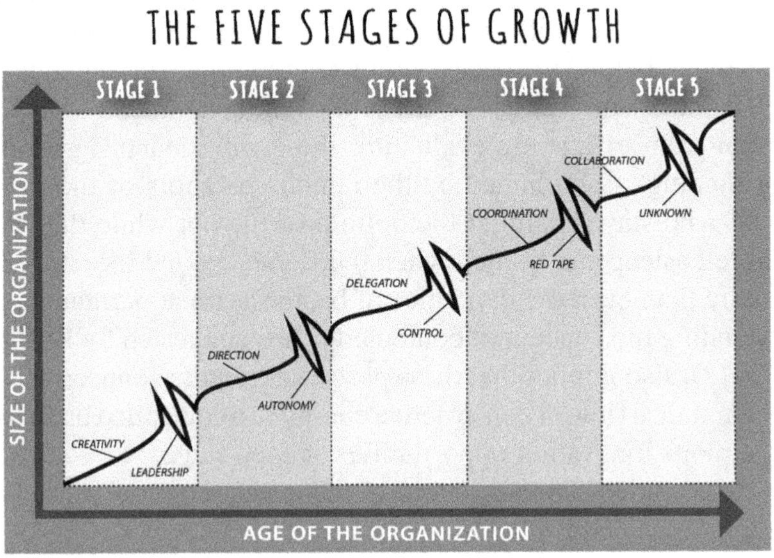

Overcoming this crisis of autonomy implies entering the next phase of growth: delegation. Once the tasks, roles and processes have been defined, human talent must align efforts.

The next stage of crisis comes when execution monitoring implies a constant supervision for the efficient performance of the organization. Controlling the performance of each participant in the process will not be a minor complication.

Establishing quality and continuous improvement programs focused on the permanent measurement of key performance indicators (KPI's), will be the tool to advance to the next stage of growth: coordination.

An efficient stage of coordination and management of a lean operation, however aberrant it may seem, will provide the foundation for the next crisis, bureaucracy, because of the lack of a holistic view generated by the specialization in single tasks.

The way to face bureaucracy will always be the collaborative work, sensitive to the needs of each customer that attends the organization, whether external or internal. The sensitivity to the processes that lives in other areas will then be the empathic solution to transcend organizationally.

However, here is where one of the most important problems for an evolving organization reside: the next crisis of innovation. In general, an organization that has achieved the highest levels of organizational efficiency tends to mistakenly believing that it can remain in this cycle permanently; assuming that the formula that led to success will keep the organization at permanent success. This is a state we called myopia of innovation. This prevents the organization from efficiently evaluating the evolutionary signals that the market is sending, as explained by the famous parabola of the frog in the boiling pot, who stayed calm at the bottom of the pot while the water was warm, but fell asleep and burned when the flames boiled the water.

Now more than ever, the dynamics of business are at permanent change, so understanding the concept of contented chaos suggested by Brafman and Pollack (2013), also implies that these processes require some creative stress that help maintain a flow of constructive energy in a controlled but permanent way that exploits innovation opportunities at each stage.

Failure to consider every single one of these stages of crisis and evolution, and to apply agile innovation strategies in each one of the tasks that an organization carries out represents a huge risk of disappearing due to lack of vision of strategic adaptation.

Figure 15.
Source: Moreno, T. (2017), Adapted from: Greiner, Larry E. (1998). Organizational Practices in The Five Phases of Growth.

ORGANIZATIONAL PRACTICES IN THE FIVE STAGES OF GROWTH

CATEGORY	STAGE 1	STAGE 2	STAGE 3	STAGE 4	STAGE 5
MANAGEMENT FOCUS	MAKE AND SELL	EFFICIENCY OF OPERATIONS	EXPASION OF MARKETS	CONSOLIDATION	PROBLEM-SOLVING AND INNOVATION
ORGANIZATIONAL STRUCTURE	INFORMAL	CENTRALIZED AND FUNCTIONAL	DECENTRALIZED AND GEOGRAPHICAL	LINE STAFF AN PRODUCT GROUPS	MATRIX OF TEAMS
TOP MANAGEMENT STYLE	INDIVIDUALISTIC AND ENTERPRENEURIAL	DIRECTIVE	DELEGATIVE	WATCHDOG	PARTICIPATIVE
CONTROL SYSTEM	MARKET RESULTS	STANDARDS AND COSTS CENTERS	REPORTS AND PROFIT CENTERS	PLANS AND INVESTMENT CENTERS	MUTUAL GOAL SETTING
MANAGEMENT REWARD EMPHASIS	OWNERSHIP	SLARY AND MERIT INCREASES	INDIVIDUAL BONUS	PROFIT SHARING AND STOCK OPTIONS	TEAM BONUS

STEP 3. EXPLORE OPPORTUNITIES: MEGATRENDS

Give me a big enough lever and foothold and I'll move the world. –*Archimedes.*

As seen throughout this first chapter, the focus has been on a process of theory and model compilations. Getting the information is the first step, filtering the information is the second step, synthesizing the information so that we can distinguish the critical aspects to address is the last step of the first stage, the stage of analysis of the situation.

Continuing with the same dynamic and addressing the topic of *megatrends*, we introduce here the concepts of one of the texts that have had greater significance in recent years in the processes of transformation, social, technological, and economic change in the society of multiple revolutions: from cheap and intense labor, to mind of value; from traditional energy resources, to self-sustaining and environmental-friendly ones; from raw material, to nanomaterials; from communication controlled by the media, to communication controlled by the users.

The book *Funky Business* (2007) by Jonas Ridderstrale and Kjell Nordstrom, sets a guideline in the traditional approach of business literature by beginning to change the traditional and scientific language and offering a different vision leading to an intellectual revolution.

Ridderstrale and Nordstrom demonstrated and suggested the process of transformation that was happening to companies at a moment where mass production was giving way to work environments open to thought and later, to collaborative innovation.

They were the first authors to recognize that management processes should be more agile and open to proposals for innovation in a collective and collaborative way. They understood that the globalization of markets was accompanied by an infinite flow of information that shifted the competition from being local to global. Now that the main competitor had access to world-class suppliers, the elements of differentiation were no longer noticeable and the competitive advantage became standardized in the market. By the way, this concept was explained in an excellent way by Kim and Mauborgne in the book *The Blue Ocean Strategy* (2005).

As previously discussed, in the case of Ciudad Juárez, the problem was not only a security matter, but also co-dependency to the American economy, as well as the myopia of innovation (to prevail in past success and comfort zones), evidencing a lack of structure and public policy that could have been focused on increasing patents and value propositions that generate different cycles of innovation.

A good part of the business sector was looking for answers by *"doing more of the same"*: Rent industrial warehouses, build, and sell more houses (for a population that was leaving the city at that time), and open more restaurants. Without even wanting to think about it, the city itself was pouring more water into the *Red Ocean.*

To fully address the subject of megatrends, Ridderstrale and Kjell Nordstrom identified the following forces:

- **The Technology:** The Digital Information
- **The Institutions:**
    - The Capitalism
    - The Nation State
    - The Political Parties
    - The Eternal Company
    - The Family
- **The Values:**
    - Work Values
    - Global Values
    - The Fusion of Values
    - The Spiritual Void

Far from wanting to explain the index of the book of Ridderstrale and Kjell Nordstrom, the goal is to present a mental map that can be compared with other similar proposals.

For example, if we followed the same exercise with the book *The 32 Trends of Change* (2010) by Juan Freire and Antonio Gutiérrez, we would find many similarities. The categorization used by them is the following:

- **Technological Transformations:**
    - Computational Trends
    - Personal Manufacturing
    - Augmented Reality
    - The Digital Body, the Digital Brain
- **Political Transformations:**
    - The New Leaderships
    - The End of the Political Community Grouping
    - Activism as an Element of Cohesion

    - Political Creativity
    - The Recovery of Words and Ideas
- **Information Transformation:**
    - Evolution of Storage Capacity and Data Reinforcement
    - The Revolution of Personal Data
    - The Information and Communication revolution
    - Mainstream Media and Grassroots Media
    - Shorter News Cycles
- **Transformations in Business Models:**
    - Access, Price, and Value Chain
    - The Paradigm Shift of Intellectual Property
    - Open Innovation Networks
    - The New Concept of Business Organization
- **Transformations in Organizations:**
    - New Working Models
    - New Models of Management: Company, Skills, and Social Networks
    - Segmentation and Social Groups
    - Corporate Website Breakdown
    - Crowdsourcing
- **Transformations in Relationships:**
    - Geolocation
    - The Third Space
    - New Privacy
    - Languages and Communication
- **Urban Transformations:**
    - Urban Computing
    - Local Citizen Participation
    - Emerging Urbanism
    - Cities as New Political and Economic Actors
    - Micro Clusters

In addition, one of our favorite materials related to strategic vision is *Abundance* (2012) by Peter Diamandis and Steven Kotler in which the challenge of "*creating a world of possibilities*" is raised.

The masterful approach of the authors not only analyzes the forces that are managing the global economy, but also brilliantly propose the opportunity to migrate towards a concept of human philosophy based on the idea of linking technology and human talent to build decent models of life for all the species that inhabit this planet.

Diamandis and Kotler, express a sense of urgency toward the need for change. They invite us to see beyond what our eyes can perceive; to observe opportunity in every problem we face. However, in order to continue the dynamics of this megatrend sub-theme, we will try to categorize its proposals in a synthetic way so that we can contrast the forces identified by the authors. In an extremely summarized way, for the *Abundance* approach (for the sole purpose of illustrating a mental map and without disrespecting the great work of its authors), we would like to focus on what they define as the base of the pyramid:

Provide sufficient water, food, and shelter.

As discussed above, the issues at the base of the Maslow's pyramid that relate to physiological and safety needs are also identified. The other levels of the pyramid have to do with needs for psychological stability.

To put this proposal into focus, Diamandis and Kotler emphasize the idea that, *"in solving our concerns about water, we will also be solving the problems of hunger, poverty, population health, and biosphere care."* (Diamandis & Kotler, 2012).

They then suggest that the real challenge is centered on the production of water for human consumption. From there, other great challenges are derived:

- Education
- Energy
- Information/Communication
- Health Care
- Social Freedom

Although they suggest from the initial trilogy that the next force with the greatest impact on the change of the global economy is energy, it is difficult for us to assert such a proposal, since we would enter the eternal dilemma of, *what came first, the egg or the hen?* (Although it has already been proven that it was the egg).

Now, although it is true that the production of water and energy that would meet the basic human needs are indisputably necessary, it is also true that in order to reach those levels of growth and development we require a mix of technology, information, and development of the human talent that accompany these priority elements. And, in developing them, we will naturally open the door to complement any process of parallel human-technological development. Once the human brain opens itself up to growth processes, it opens itself, exponentially, to development processes.

However, to try to synthesize the megatrends that the authors recognize in their brilliant masterpiece, we will concentrate on listing some forces that we consider as the most remarkable:

- Biotechnology
- Computer systems
- Networks and sensors
- Artificial intelligence
- Robotics
- Digital Manufacturing
- Medicine
- Nanomaterials
- Nanotechnologies

At the same time, there are reports that are continuously monitoring opportunities and developments in different global economies. One of the most respected reports of this kind was executed by the Global Economic Forum which, in 2015, identified the top 10 world challenges:

1. Furthering of the Income Gap
2. Constant Increase of Unemployment
3. Lack of Leadership
4. Growth of Geo-Strategic Competition
5. Representative Democracy Awakening
6. Increased Pollution
7. Increased Severe Weather Events
8. Intensification of Nationalism
9. Increased Stress due to Lack of Water
10. Growth of the Importance of Health in the Economy

Additionally, from the technological trends' point of view, the 2016 report identifies the following 10 as the most relevant:

1. Nano Sensors and the Internet of Nano things
2. Next Generation Batteries

3. The Blockchain
4. 2D Materials
5. Autonomous Vehicles
6. Organs-on-chips
7. Perovskite Solar Cells
8. Open AI Ecosystem
9. Optogenetics
10. Systems Metabolic Engineering

To complement the megatrend approach, we considered necessary to include in this comparative summary the complementary view of the Latin journalist Andrés Oppenheimer.

In two of his most recognized works Oppenheimer talks about the importance of developing a strategic approach to innovation in Latin America. It is urgent that the countries that make up the Central and South American block, including Mexico, stop finding answers for the future in the past. Oppenheimer also asks us to stop reading the stories of Francisco Villa; quit "*exhuming the dead*" (pointing to the case of Simon Bolivar) to begin to understand and capitalize on the stories of a global market that has bet on science, education, and technology as the engine of development for the well-being of their population.

In his book, *Enough Stories* (2010), Oppenheimer makes a comparison of European and Asian economies against the performance and approach of Latin American countries. It explains how South Korea registers more than 7,500 patents against Brazil with only 100; revealing an obsolete and inefficient educational system in Latin America.

However, in his following publication *Create or Die* (2014), Oppenheimer complements the analysis made in *Enough of Stories*, identifying macroeconomic megatrends that suggest a course of action to the future, with direct implications in the present.

We are living in the global knowledge economy, where the fastest growing nations—also reducing their poverty—are those that produce technological innovations. Today, the prosperity of countries depends less and less on their natural resources and increasingly on their educational systems, their scientists and their innovators.(Oppenheimer, 2014)

He identifies (like Diamandis and Kotler) some trends in different aspects, most of them technological and related to education and innovation:

- 3D Printers that Make Shoes
- Drones that Deliver Pizza
- Self-driving Cars
- Auto-healing materials
- The Internet of Things
- Big Data: The Gold of the 21st Century
- Hand Watches with Heart Rate Monitor
- The Supercomputer that Prescribes Medicine
- Personalized Education
- Trips to the Stars

And of course, he talks about Diamandis and Kotler`s work:

We are entering a period of radical transformation, where technology has the potential to substantially improve the average living standards of every man, woman and child on the planet. Within a generation we will be able to give them goods and services that in the past were reserved for a few rich people who need or want them. Abundance for all is within our reach. (Diamandis, P.H & Kotler, S, 2012).

We would like to summarize that, regardless of the sources that illustrate global megatrends, there are some categories that are repeated in them:

- Informational Technologies
- Technological Development
- Health and Human Development
- Community Development
- Political System Structures Development
- Communicational Structures and Tools
- New Business Model Development

Megatrends only set the tone for the next three to five years. The dynamics of today's business do not allow us to execute strategies for longer periods of time. There will be those who, traditionally, will continue to think that a 10-year strategic plan gives will give them a clear guide to follow. However, for the authors of this book, this concept is unproductive and far from reality.

Dedicating resources to define the path of the next 10 years makes no sense when technology and new devices are replaced almost every year, as in the case of the iPhone. Some will argue that a 10-year strategic plan is useful in terms of vision; however, megatrends set a new guideline for these plans to be implemented. Therefore, we suggest adapting the term strategic planning to strategic adequacy.

To make a summary of every single prediction or proposal that exist trying to explain the forces that move the world would be of little value if we do not find the applicability in the current context of entrepreneurship. Due to changes in most the cases which live on the basis of technological advances, we must understand the objective of situational analysis, which is nothing more than providing information to facilitate an efficient decision-making process.

At the *Tecnológico de Monterrey; Monterrey Institute of Technology and Higher Education* (Technological and Higher Learning Institute of Monterrey) where we work, the definition of entrepreneurship specifies the need to "*have the passion to generate creative solutions that transform the economic, social, and environmental reality of our communities.*"

Entrepreneurship must consider every challenge and opportunity settled by megatrends across the globe and transform them into solution alternatives that impact human well-being.

Understanding the concept of "*wellbeing*" as an equation:

(Well-Educated) + (Well-Doing) + (Well-Owned) = (Well-Being)

Applied to four different contexts: the family, individual, work, and community.

STEP 4. DEVELOP THE RIGHT ATTITUDE: THE ENTREPRENEUR'S MINDSET

For more than 75 years, the Mexican University; *Tecnológico de Monterrey; Monterrey Institute of Technology and Higher Education* has been identified as an institution of higher education that promotes entrepreneurial leadership. As time goes by, the concept of entrepreneurship has taken force in the world, and the institution was in need of defining more clearly what would be the characteristics to develop in each and every one of the students who passed through its classrooms. The definition of these characteristics would help to focus the teaching strategy of such a prestigious institution without limiting the approach and, instead, prioritizing the basis of its entrepreneurship model.

The vision of the entrepreneurial leadership with a human sense began to become tangible with the definition of these four characteristics:

1. Tolerance to Risk and Uncertainty
2. Swift Recovery from Failure and Didactic Experience
3. Ability to Identify Opportunities (problems); Implementing Innovative Solutions
4. Ability to Obtain Resources to Implement Value Propositions

The development and implementation of different teaching strategies focused on learning based on the entrepreneurship model have made us understand that, in addition to these four variables there are some others (of individual and team) which complement the entrepreneur's mindset.

Some of the most relevant are:

1. **Resilience:** The ability to recover optimism repeatedly in the face of the constant impacts of the work environment.
2. **Perseverance:** The desire to "*push through*" the end without losing a drop of enthusiasm.
3. **Creative Leadership:** Mental strength to support and adapt an idea.
4. **Collaborative Work:** Unlike teamwork (that usually divides the tasks), this emphasizes the idea of complementing competences among the members.
5. **Disruptive Vision:** The capacity of the human being to see what everyone else sees and dream what others cannot understand.

Complementary literature that not only supports, but also complements the profile of an entrepreneur has been proposed by different authors. In the first case, Jeff Dyer, Hal Gregersen, and Clayton M. Christensen in their book *The Innovator's DNA* (2011), suggest four key skills in the personality of the innovator—which is not synonymous with entrepreneur but finds its similarity in the creative mindset.

In a nutshell, the skills are presented as:

- **Discovery Skill 1 - Association:** Innovators think differently because they relate what is (apparently) unrelated. The association, or the ability to make unexpected connections between different areas of knowledge, is a skill that is often taken for granted among innovators.

- **Discovery Skill 2 - Questioning:** Disruptive innovators develop their work by asking questions that invite reflection. Questioning is the creative catalyst for other discovery behaviors: observation, networking, and experimentation.
- **Discovery Skill 3 - Observation:** Most innovators are sharp observers. They look closely at the world around them and, when they observe how things work, they often identify that which is wrong.
- **Discovery Skill 4 - Networking:** Thinking creatively often involves connecting the ideas from our area of knowledge with those of people who "play in different fields," or are outside of our field of influence.
- **Discovery Skill 5 - Experimentation:** Good experimenters know that while questioning, observing, and developing contact networks that provide data on the past and the present, experimentation is the most viable way to generate information about what can function in the future.

Figure 16.
Source: Moreno, T. (2017), Adapted from: World Economic Forum. (2015). New Vision for Education. May 2017, de The Boston Consulting.

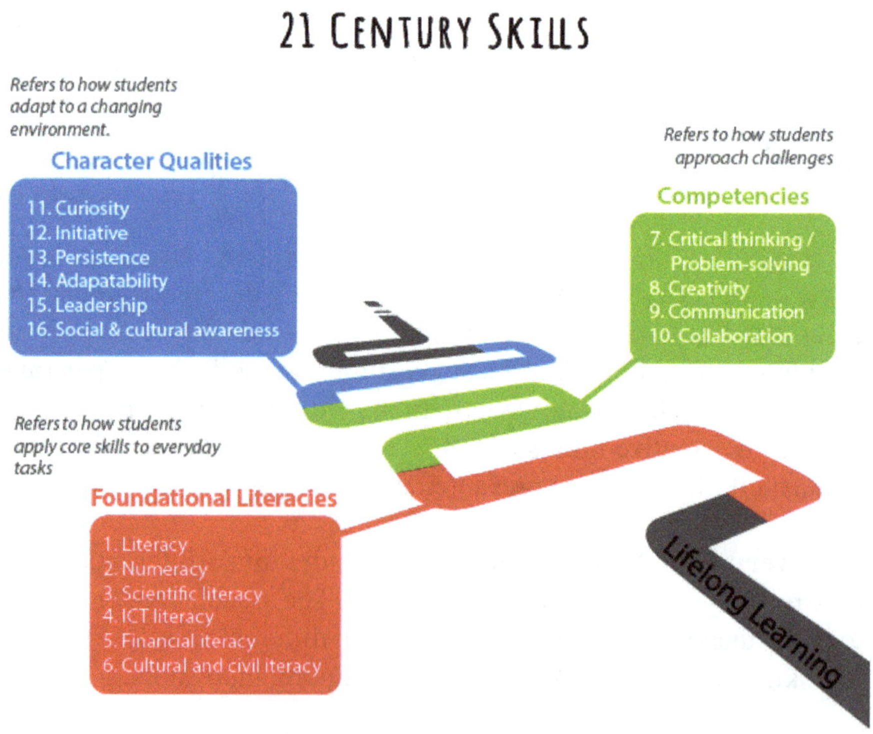

Figure 17.
Source: Moreno, T. (2017), Adapted from: Dweck 2007.

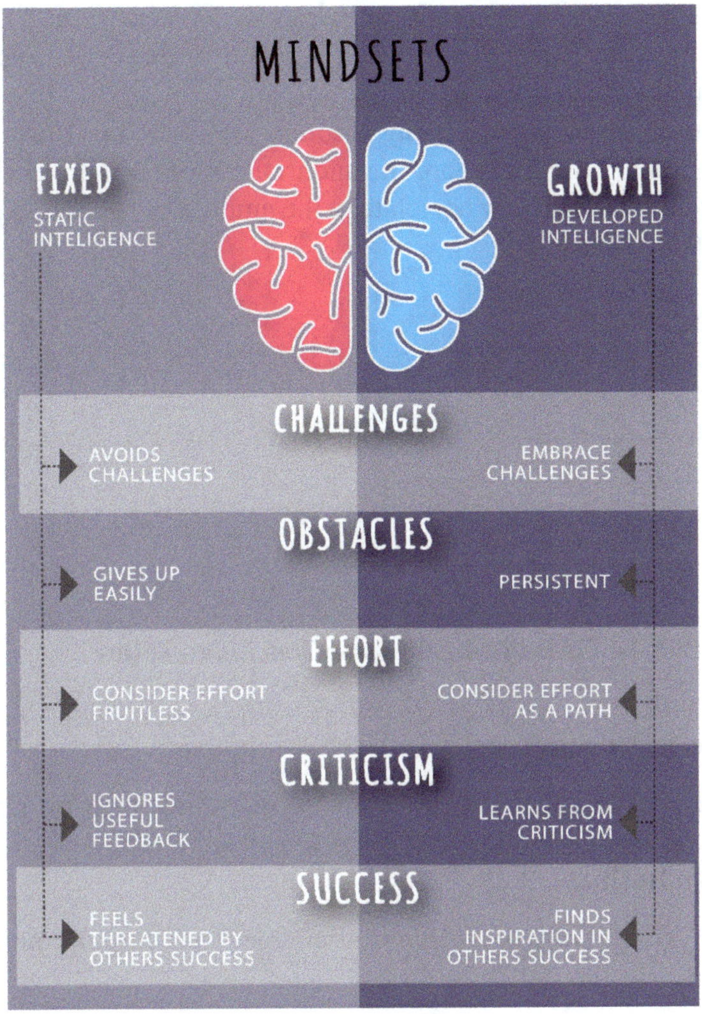

The World Economic Forum's website includes in its publication abilities that, from its point of view, are the main focuses in child development (Figure 16).

The importance of this proposal lies in understanding how the two roles: a) competencies and b) qualities of character, play in the formation of the entrepreneurial spirit from an early age.

We will conclude this topic by mentioning one of the most relevant works in the definition of competencies or skills of an entrepreneur. Carol Dweck in her book *Mindset, The New Psychology of Success* (2007) develops a comparative scheme of a fixed (static) mentality and a (entrepreneurial) growth mentality.

The attitude that a human being assumes and develops during five key elements in his/her life (or before the undertaking of each project in his life, we should add), will directly correlate to the amount of success experienced.

These elements are:

- Challenges
- Obstacles
- Effort
- Review
- Other Success

REFERENCES

Batista, E. (2007). *Ed Batista Executive Coaching*. Author.

Blueoceanstrategy.com. (2005). *Blue Ocean Strategy*. Available from: https://www.blueoceanstrategy.com/

Brafman, O., & Pollack, J. (2013). *The Chaos Imperative: How Chance and Disruption Increase Innovation, Effectiveness, and Success* (1st ed.). New York: Crown Business.

Bruton, A. (2016). *The Innographer.* Available from: http://theinnographer.com/about

Bucay, J (2013). *De la ignorancia a la sabiduría: El camino de Shimriti.* Océano exprés.

Diamandis, P. H., & Kotler, S. (2012). Abundance: The Future Is Better Than You Think. Academic Press.

Dweck, C. S. (2007). *Mindset: The New Psychology of Success* (1st ed.). New York: Ballantine Books.

Dyer, J. (2011). *The Innovator's DNA: Mastering the Five Skills of Disruptive Innovators*. Library of Congress.

Frankl, V. E. (1946). *Man's search for meaning* (1st ed.). Beacon Press.

FreireJ.GutierrezA. (2010). *EUAbe*. Available from: http://www.eua.be/Libraries/higher-education/trends2010.pdf?sfvrsn=0

Garcia Marquez, G. (2007). *Love in the Time of Cholera*. Academic Press.

Greiner, L. E. (1998). Evolution and Revolution in the Growth of Organizations. *Harvard Business Review*.

Hbrorg. (1998). *Harvard Business Review*. Available from: https://hbr.org/1998/05/evolution-and-revolution-as-organizations-grow

Jampolsky, G. (2003). *Love is Letting Go of Fear* (1st ed.). Celestial Arts.

Jodorowsky, A. (2012). *Ojo de Oro: Metaforismos, Psicoproverbios y Poesofía*. Editorial Siruela.

Jung, C. (1928). *Contributions to Analytical Psychology*. Academic Press.

Kawasaki, G. (2004). *The Art of the Start: The Time-Tested, Battle-Hardened Guide for Anyone Starting Anything*. Penguin Group.

Kolb, D. (2017). *Identainet*. Available from: https://ebls.identai.net/about/

Kotter, J. (1996). *Leading Change* (1st ed.). Harvard Business School Press.

Kumar, V. (2012). *101 Design Methods: A Structured Approach for Driving Innovation in Your Organization* (1st ed.). John Wiley & Sons Inc.

Managementstudyguide.com. (2017). Available from: http://www.managementstudyguide.com/maslows-hierarchy-needs-theory.htm

Mauraya, A (2012). *Lean Canvas*. Academic Press.

Oppenheimer, A. (2010). *Enough stories*. Debate Editorial.

Oppenheimer, A. (2014). *Create or Die*. Vintage Español.

Osterwalder, A. (2010). *Business Model Generation: A Handbook for Visionaries, Game Changers, and Challengers* (1st ed.). Wiley.

Peter, S. (1990). *The Fifth Discipline*. The Art & Practice of the Learning Organization.

Portillo, E. (2016). *Advanced Technologies Research and Integration Center*. CIITA.

Portillo, E. (2016). *Triple Helix Integration Model*. Academic Press.

Ridderstrale, J., & Nordstrom, K. (2007). *Funky Business Forever: How to enjoy capitalism*. Prentice Hall.

Schon, D. (1992). *La formacion de profesionales reflexivos: hacia un nuevo diseño d e la enseñanza y el aprendizaje en las profesiones* (2nd ed.). PAIDOS IBERICA.

The Arbinger Institute. (2010). *Leadership and Self-Deception: Getting Out of the Box*. Berrett-Koehler Publishers, Inc.

Trainer-tools.com. (2015). *Trainer Tools*. Available from: https://trainer-tools.com/contribute/roger-greenaway/

Valenzuela, A. (2014). *Conversation with "Mesa de Seguridad de Ciudad Juárez"*. Academic Press.

World Economic Forum. (2015). *New Vision for Education*. The Boston Consulting.

KEY TERMS AND DEFINITIONS

DNA Definition: The "why" factor or motive that leads to the creation of a business/company/enterprise.

Incubation Stage: Period of validation for a business model.

Maquiladora Industry: A factory in Mexico run by a foreign company that exports its products to the company's country of origin.

Mega Trends: Long-term change that affects governments, societies, and economies permanently over a long period of time.

Red Ocean: Simile on *Blue Ocean Strategy* referring to an industry which is overwhelmed in its market.

Silos: Structure used to store grain. In marketing, simile portraying an organization in which the work is done individually, "burying" themselves in their area without acknowledging the necessity of working as a team/ecosystem.

Spinoff: For the purpose of this book, spinoffs are defined as corporate entrepreneurship initiatives.

White Elephant: Public construction project; maintenance or installation of public property which results in a negative impact on the community due to lack of completion or—in case of successful completion—lack of revenue.

Chapter 2
The Definition of DNA

ABSTRACT

The importance of defining a central purpose (or DNA as we call it in this book) is to have an idea, as clear as possible, of what motivates us in life (personal and corporate). Knowing who we are and where we want to go is the first step, as we saw in the previous chapter. However, when we talk about agile corporate innovation, the first step is to define a guiding team: a small group of 4-5 people who start with the work of strategic focus for the company, in whose number lies precisely the strength of the agile approach. This group of leaders needs to build a vision that later leads to the generation of an organizational structure that allows achieving the mega goals and objectives set for each proposed innovation initiative. The selection of the appropriate human talent and the execution and correct measurement of the proposed actions are the third and fourth steps. This chapter explores the authors' definition of DNA.

STEP 5. THE IMPORTANCE OF A COLLABORATIVE APPROACH: THE GUIDANCE TEAM

To place ourselves in the process, let's review the Stragile Model one more time (Figure 1).

A state of awareness, as explained in chapter one, acts like the trigger for any process of evolution at individual, family, labor, and community levels. To define as clearly as possible who I am, at what point in life I am standing, in which place, before what opportunities and what attitudes skills

DOI: 10.4018/978-1-5225-4023-6.ch002

Figure 1.
Source: Portillo, E. (2016), The Stragile Conscious Corporate Innovation Model.

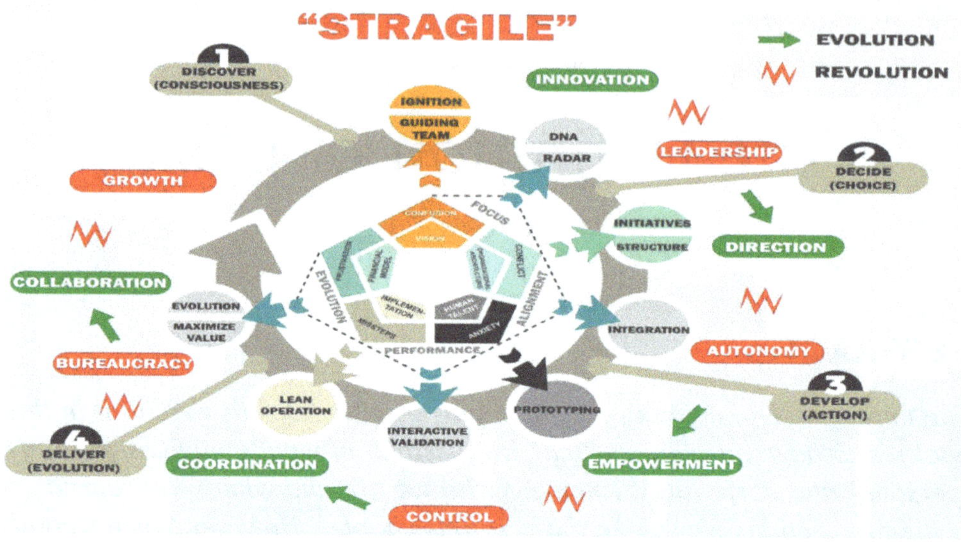

and competencies I require to face those opportunities; the state of awareness is *definitely* the first step.

Once the ignition process has taken place, the following reflection is: *What do I need to turn my idea into a reality?*

There are two ways: Build it with or without help.

The capacity for partnership we have will be one of our main weapons to achieve our goals in the short term.

A lone wolf does not hunt big prey.

It is often seen that individuals who live moments of inspiration (or vision, properly speaking) are more prone to generate creative alternative solutions to the every day problems. Also, much more frequently, individuals tend to protect "*their great ideas*" resulting in concepts being buried deep into the darkest part of the brain.

Experience tells us that an idea that is not shared, validated, and constructed, is only an illusion.

Creative thinking must be transformed into innovation and, most importantly, must be marketed! We cannot talk about entrepreneurship when the cycle stays still at the stage of the idea, just as we cannot talk about entrepreneurship if innovation stays at the stage of invention.

An invention that does not *"live"* or goes through the sales process (not only from the economic point of view) is destined to become a museum piece, at best.

Very often we witness discussions between friends or colleagues who feel that someone *"stole their idea,"* or that by proposing the idea they feel etitled to the highest percentage of the company because of their wrongful belief of contribution. Wanting upwards of 70% of the company sometimes, without realizing that moving from a creative idea to a successful business requires energy and time in order to materialize a dream. Most of the time, after withstanding market validation, the initial dream ends up being a completely different concept to what it was destined in its earliest stage.

Kotter's model can better exemplify the basis of this stage's proposal (Figure 2).

A sense of urgency must be provided by the reflections caused from the approach of chapter one: the motivation, megatrends, and DNA of the entrepreneur.

The formation of a guidance team is the next step proposed by Kotter to ignite the organizational change—although the same applies to the formation of a new company. We would like to emphasize in detail the application of this model under the strategic adequacy approach: the guidance team is composed of several elements. However, what is fascinating in the practical application

Figure 2.
Source: Moreno, T. (2017), Adapted from: Richman, R. (2015).

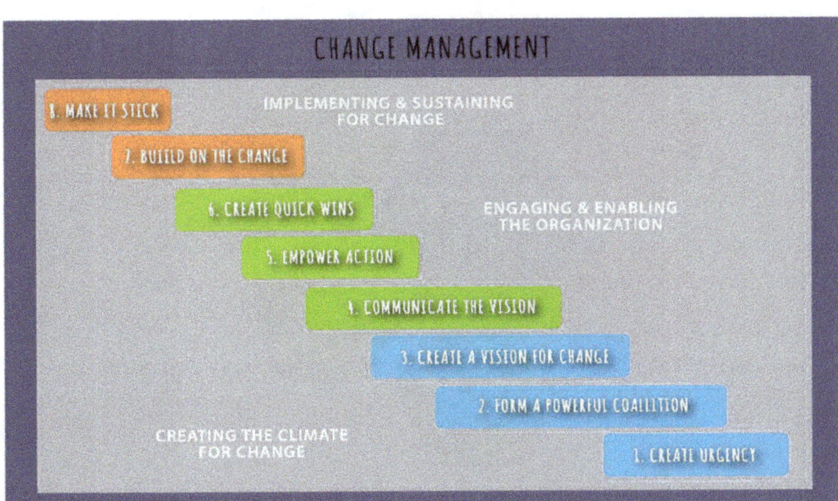

of this methodology during the launch of corporate innovation initiatives (or companies in-the-making) is that the base of a guidance team shall not have more than 5 people with *real decision-making power* at its definition stage.

What does this mean? Before the appearance of an innovation initiative, entrepreneurship, improvement, reengineering, redesign, organizational change, or however the term of innovation applies best, the initial teams (the executives) are faced with a real dilemma: Decisions (often with clashing egoes), and will suffer the consequences of the Quantity VS. Quality. At this start-up phase, more is not necessarily better.

A small number of decision makers (Executive Guide Team) will be able to define with greater ease the first innovation proposals to be developed. As the team evolves its idea, the integration of the rest of the guide team's elements will change gradually. After the Executive Guide Team comes the defining process of who the Base Team and the Extended Team will be.

Figure 3 shows the scheme of how a Leading Team must be formed. Once defined the Executive Team (represented in this example by the institution's

Figure 3.
Source: Portillo, E. Moreno, T. (2017), Stragile Conscious Corporate Innovation. Adapted from: Tecnológico de Monterrey; Monterrey Institute of Technology and Higher Education Leading for Change Model.

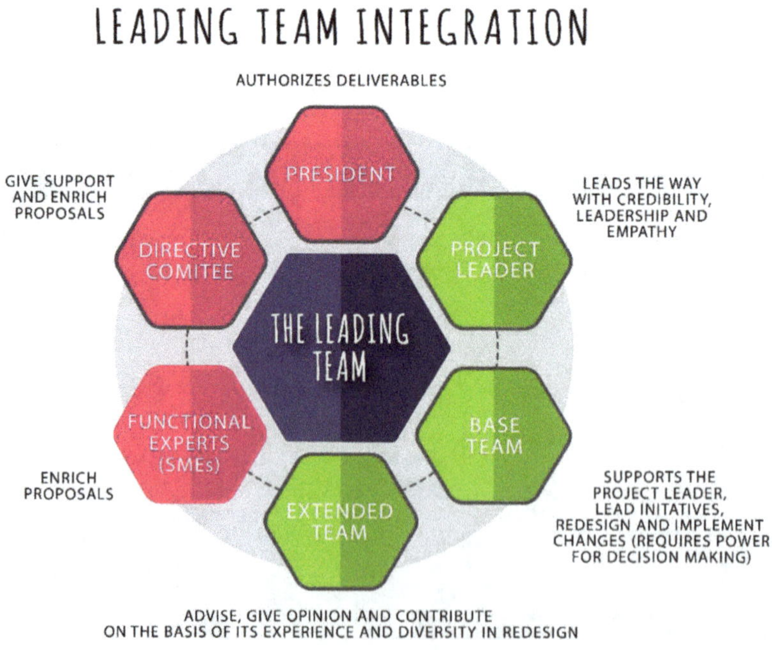

board of directors or Directive Committee), the next step is to work on the integration of the Strategic Team. Within this Strategic Team are the President (represented in this example under the figure of a Rector) and a team of external strategic advisors or functional experts. At the same time, the person responsible for implementing (operating) the action plan (usually represented by the project leader) must be identified, as well.

Accuracy is necessary at this point. It is recommended to replace the term "*process owner*" to: project leader or director. This observation was shared by the leader of one of the most important manufacturing companies of the Paso del Norte region (Ciudad Juárez Chih., El Paso Tx., Las Cruces NM). He commented,

We have experienced in our company that, by using this term [owner], some of the people who assume this responsibility end up backfiring. For them, the word "owner" gives them a sense of omnipotence and empowerment that sooner or later creates conflict with their teams.

This suggestion weights on the idea of adjusting the term to *Leader*.

Once the base of operation for the *Leading Team* has been defined, the next step is to determine the innovation initiatives necessary for the construction of the organizational structure. We must define, clearly at this point, what we need to do in order to reach the ideal state defined previously.

From the perspective of the Stragile Model, the next step, after the definition, is the integration of the personnel participating in the formation of a *Base Team* by their own initiative, as well as the definition of an *Extended Team* that support the development of each initiative.

Key elements in the conformation of the Leading Teams before the innovation management.

At this point, we must take a bit more seriously Peter Senge's proposal according to the value of the 5 disciplines of an intelligent organization (Figure 4).

Human talent is the ultimate driving force of any organization; a group of human beings who share a vision that is also aligned with their personal purposes turns a profession into a vocation. The expression, "*I would do it even if they didn't pay me*" has a connotation of great relevance at this point. It means that what is being done professionally holds a strong value for the individual; enough to consider his work a delight.

Personal domain allows whoever develops it, to reach a state of consciousness and maturity such that he/she stops worrying about being fired

Figure 4.
Source: Moreno, T. (2017), Adapted from: Senge, P. (1990).

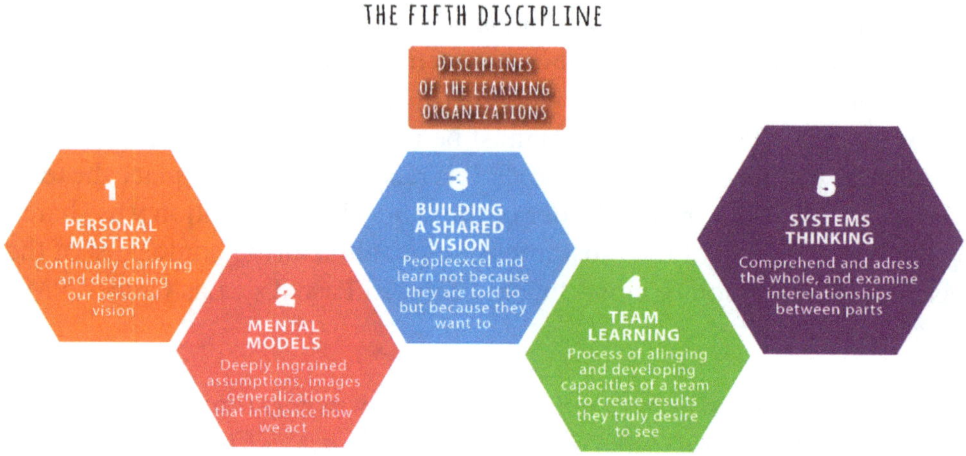

and concentrates on adding value to the tasks he/she performs. This results in a positive outcome since the team member no longer stresses on keeping his job, but in contributing to achieve organizational goals.

Everything that annoys you from other beings is nothing more than a self-proyection of unresolved issues. Siddharta Gautama Buda

The individual who reaches this level of maturity and understands the systemic approach, stops thinking about divisions; instead thinks and defends a single entity. They cease to be affect by power struggles and positions of arrogance, or protagonists that tend to flatter bosses and usually identify these postures as signs of immaturity. Anyone who reaches this level of dominance, for the most part, does not live concerned about the budget. On the contrary, always considers the way to attract external resources to the organization and achieve its objectives.

An individual, in the *personal domain* state, understands that it has lived in crisis since it has had use of reason, and that, therefore, crisis rather than situations of panic generates opportunities for growth. This is how we can differentiate individuals with high resistance to change, who constantly repeat why new ideas will never prosper, from those whose preferred questions like: *how can we make this work*? The latter spend their time wondering and imagining: *How could you better what you are already doing*? People who reach this level of domination will never know the concept of comfort zone.

On the contrary, they adopt change as their comfort zone. They do not know resistance to change and crisis became part of their evolutionary process, both personal and institutional. As Senge explains, their daily nourishment is creative tension (Senge, 1990).

As part of the talks and interventions we have had with some companies advised by us, we understood that mental models were strongly linked to an organization's leadership style. Thus, we find in this "*walk*" of corporate entrepreneurship, different postures before the diverse challenges that were divided onto each organization.

For example, when we talked to a manager of a major manufacturing company about innovation issues, the answer was:

Let it go, that issue is not well accepted here, innovation is a matter of the design center at the United States. We do not need creative minds; we cannot distract ourselves from the processes of quality control that allow us to be a world-class company. Our goal is manufacturing, not designing.

This is a clear example of a mental model that reflects an organizational culture. As long as this company continues to have that leadership approach, it can never evolve to the next level.

On a different occasion, the security crisis led us to interact with another businessman who was on the verge of despair because his business was suffering the consequences of lack of security. We discussed innovation, developing new conceptsl as well as local, regional, and international growth, but in the end, after having been working from a corporate approach, the definition of its DNA, as well as the strategic initiatives that would ignite its growth, focused on renewing business concepts that had generated success in the past and reinforcing them with the most basic marketing tool to attract traffic to businesses: sales promotion.

I must acknowledge that my inability to transform the mental model of the organization leader led us to keep doing more of the same while all the staff was ready and motivated to move the next step. Although, I must also acknowledge that, perhaps, what I did not understand at that moment, as in the parable of the frog and the scorpion, is that the scorpion sooner or later tends to do what he has always done by instinct: inject his poison. Today, the Stragile model centered on the four stages: *awareness*, *choice*, *action*, and *evolution*, invites leaders to see beyond what's in front of them; to take responsibility for making decisions and act based on what they can see. We prefer to act and learn fast in order to evolve faster.

The third anecdote we want to share regarding the subject of mental models talks about a leader who changes his mind every week. The company was constantly living in theory; a process of transition. Even the leader had decided not to operate in the office of his company. Instead, he moved the office to the living room of his own house. Due to that level of indecisiveness, the whole team of managers had to travel every day to the employer's house to work in the model of strategic adaptation.

However, such was the motivation of the team before the opportunity of change that the unthinkable happened, the *Extended Teams* became so committed to achieving goals that set off a series of highly attractive strategies. The level of motivation was such that even a job offer was extended to me. A week later, the strategic alignment project was canceled. The reason? The organization did not have the maturity nor was prepared for a new stage.

The real reason? The businessman felt threatened by the possibility of being displaced from his leadership position before the staff. Later, we found out two things: some people lost their jobs due to dismissal or health issues (the most serious case being a hospitalization of several days), and the second was that most of the strategies proposed were finally applied several months later under the leader's initiative.

Finally, one of the most emblematic cases that we have had the pleasure to work with, was that of a company whose branch in this city was about to disappear. Today, with much more experience from every one of the projects that we have advised, we can conclude in this first phase of the application of the Stragile Model, that the key behind it all relies on the type of leadership you want for your business. The general manager was worried that the low sales were drowning the company; the referred crisis did not allow traditional sales strategies. However, the type of leadership he assumed was open to change. Every day he had more questions for us (his advisors) and for his team.

He set aside the intimidational leadership (common in several cases of the manufacturing industry) and replaced it with informational leadership, but specially with *inclusion* and *inspiration*. This manager was able to combine each and every one of the tools available to lead the change and associate them with the tools presented in the Stragile methodology. The greatest satisfaction of working with this person was to be able to witness this level of wisdom and intelligence, and remembering a phrase that sets apart the wise from the fool:

The whole problem with the world is that fools and fanatics are always so certain of themselves, but wiser people so full of doubts. (Russell, 1933)

Figure 5.
Source: Moreno, T. (2017), Adapted from: Denning, S. (2011). How Do You Change An Organizational Culture?, FORBES.

This leadership comparative was perhaps one of the most enriching lessons to determine the strategy to follow when selecting new projects.

Regarding the discipline of team learning, in –addition to the observations, definitions and recommendations of the book *The Fifth Discipline*, we want to add a combination of elements that we consider to be key in the process of integration and consolidation of a team (Figure 6).

This model presents us with 6 different variables, for which we will try to explain as clearly as possible the way they interact as well as the relevance and impact they have during the formation of solid work teams.

1. **Communication:** Communication has become the most important tool for the evolution of the human being. Not only the man has the ability to communicate, other animals such as: sean birds, wolves, or elephants have an impressive level of communication. However, it has been the man who has been able to integrate thought and verbal communication. That single trait of intelligence, although some of us doubt that intelligence

Figure 6.
Source: Portillo, E. & Moreno, T. (2016), The Stragile Cycle for building loyalty.

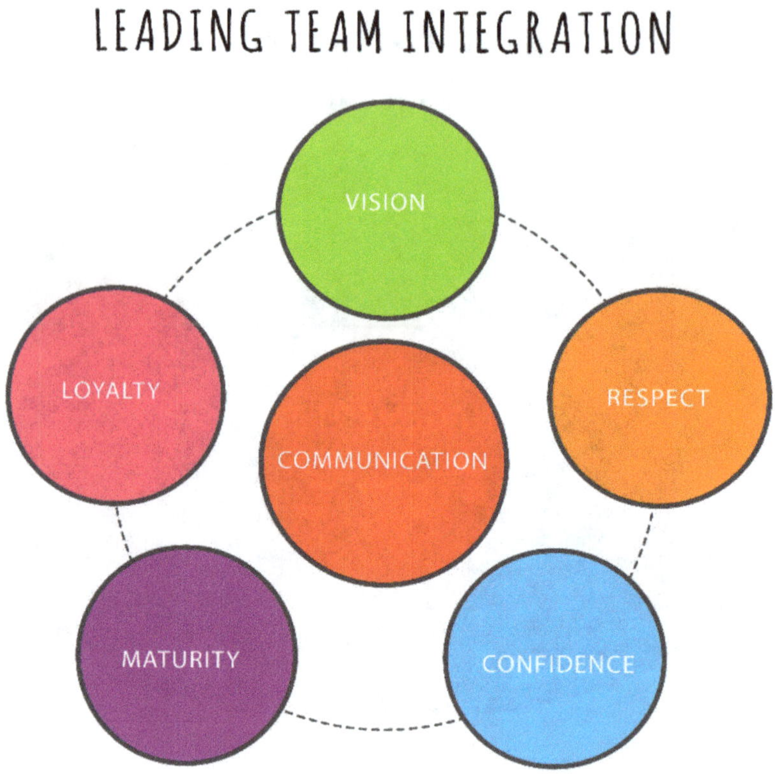

is an exclusive trait of humans, has helped communication evolve into networks, and later into learning. That communication becomes innovation that turns into entrepreneurship. The power of communication transcends paradigms, impacts individual and collective growth and evolution. When we close to communication, we close the door to the opportunity for learning and development. Communication is then the basis of human progress.

2. **Vision:** Under the Stragile Model, the concept of vision implies a blend of the individual vision—suggested in the concepts of personal and shared vision that Senge illustrates. As we have been insisting, every process of transformation begins with a state of awareness: who I am and where I want to go. The vision then represents perhaps the most exhaustive effort during the growth phase of an innovation project. Defining a purpose at an individual level is already complicated, now imagine the complexities of aligning an entire organization. It is inevitable at this

point to remember Nemo. When he tried to coordinate the individual efforts, and focus the tunas to direct them to the bottom of the sea. The strength of a disoriented mass was useless, as they swam from one side to the other in a futile effort to reach the goal of being saved from the nets. It was the vision, the commitment, and the determination of the small fish, that managed to turn all the fish in the right direction.

3. **Respect:** For oneself and also for others. At an individual level, it implies to have a little more appreciation for ourselves and to recognize that, although it is true that we are not perfect, it is also true that as human beings, we have or we can develop the necessary skills to complement our collaborators. It would be worth resuming the approach of Buckingham and Clifton, in his book *Now Discover your Strengths*, where the central thesis orbits around the idea that we humans tend to be perfectionists, which makes us focus (and with more effort than necessary) in our imperfections. The suggestion of the authors proposed concentration on what we are really passionate about and above all, that we do well. It is our strength the sole responsibe for guiding our talent. Learning to respect ourselves is then the first step; the second, respect others. The word respect implies consideration, admiration, listening, attention; a way of appreciating the essence of other people or opinions without venerating one's own. Respect for the views and participation of others in a work team is essential to build strong human connections. Understanding who you are and respecting other's opinions is key to build trust. It is necessary to remember at this point the Dunning-Kruger effect, which states that people with low intellectual level tend to think they are smarter than they are. (*The Journal of Personality and Social Psychology* by Justin Krugger and David Dunning, 1999). For the purposes of this book means to consider the fact that a fool usually speaks from the security that ignorance gives him. However, for true respect to work "*it helps to remember that each person is doing their best from their own level of consciousness*" -Deepak Chopra.
4. **Trust:** It implies having blind faith in that our counterpart, collaborator, partner, friend, relative, boss, or employee will impeccably develop the part that corresponds to him. It infers to have enough character to not hesitate when the circumstances, the facts, and the people could be showing a possible failure. But we must be careful in this delicate matter, when a person approaches you wanting to make a complaint against a person you trust, our recommendation is:

a. If you do not have anything good to say, do not say anything.
b. Avoid talking and letting people speak badly about others who are not around at the time. When the situation really requires it, interrupt the session and ask the person in question to be present to discuss the matter.

5. **Maturity:** Another tricky subject to tackle from a single perspective. In general, maturity is interpreted as an emotional state in which a person has ceased to show signs of childish behavior. The perspective from which we want to aboard this variable is precisely from the point of view of emotional intelligence. Human coexistence is complicated when a state of immaturity (also associated with a state of fear of behaviors or events that we suppose can result in some kind of damage) is present. Emotional intelligence is perhaps of far greater relevance than logical-mathematical intelligence. Of course, this statement implies a comparison and requires a validation under rigorous criteria which, given the variables, is complicated from a scientific perspective. It is also possible for someone to think it depends on the goal to be achieved. However, under Stragile perspective in building highly efficient teams, emotional intelligence makes a lot more sense. It includes intrapersonal intelligence (ability to understand themselves) and interpersonal intelligence (ability to get along with others). It consists of five key elements: self-control, self-awareness, self-motivation, empathy, and social skills. A person with a mathematical logic that does not have the ability to self-control, self-motivate, or socially interact, will face the challenge of transcendence, since it will tend to isolation—generated by him or by others.
6. **Loyalty:** Reaching this level in a team involves a high degree of communication, respect, reciprocity, commitment, and gratitude from one person to another. Loyalty is the highest degree of expression of fidelity. It means that you do not experience any credibility doubts about another person. It implies temper and character but, above all, a high sense of respect. Loyalty sometimes reaches levels of surrender and sacrifice such, that whoever lives it would not hesitate a second to do what the team requires to achieve the collective goal. Seen as such, it is easy to think that loyalty should be the main driver in the collaborative work approach, but it may not necessarily be true or it may have some limits; especially when loyalty to the team or task goes beyond the holistic or systemic approach of an organization. Loyalty to the vision should never be confused with loyalty to individuals.

In the shared vision approach, a team's ability to see "*the entire film* instead of *just a frame*" has a high impact on their ability to integrate cohesively and productively. To conceptualize this, we will share the adaptation of an image that appeared on the internet and from which, we lack an attributable source (Figure 7).

To talk about *shared vision* implies a collaborative approach, communication, coordination, interconnected human and organizational links in such a way that the constant, orderly and correct flow of energy, allow organisms to operate in the most efficient way possible. We insist to explaining what Senge has so brilliantly exposed, the book *The Fifth Discipline* allows us to present an idea that frames our holistic view of personal, family, work, and community cycles. The beginning of each cycle places us in acting unconsciously wrong. Usually our mental maps allow us to see reality in a limited way. However, in our unconscious ignorance, we tend to think that what we do or say is the right thing, without realizing that we are acting wrong.

Figure 7.
Source: Portillo, E. & Moreno, T. (2017), The Stragile Conscious Phases to Entrepreneurship Culture Development.

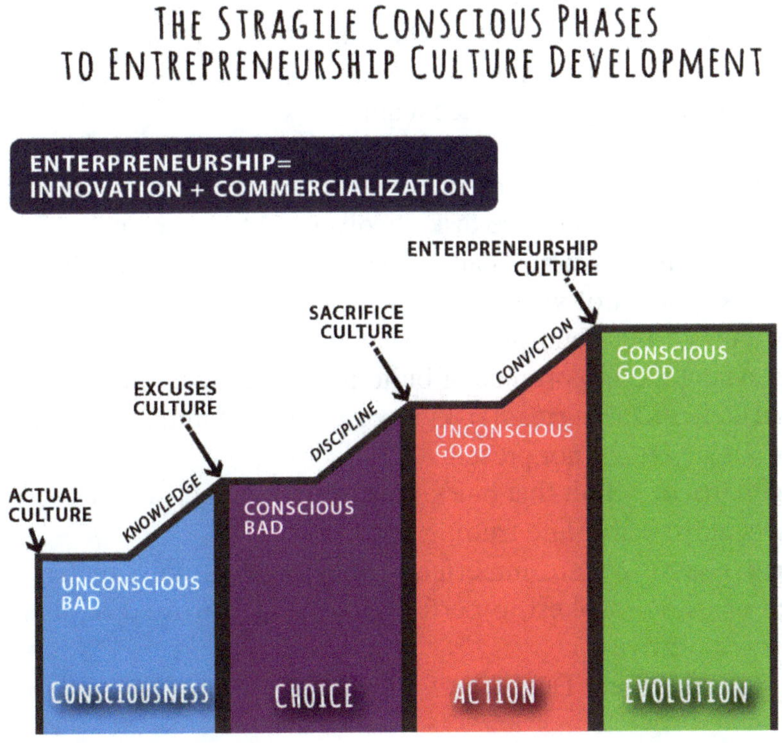

The way to the next level of consciousness is through knowledge. Better understanding of the situation where we stand and the variables that come into play gives us the opportunity to make informed decisions.

However, knowledge only leads us to a stage where we continue to act poorly but consciously. This stage is defined as a stage of pretexts or justifications. To protect or disguise our ineptitude we will blame others for the failures in the organization. Discipline in the practice of our tasks will lead to specialization and dominion of tasks. Only then we will then have reached the stage of acting positively and consciously.

It would seem that, after awareness and mastery of one's tasks, we would have reached our goal; however, a step is missing. To act well unconsciously, it is necessary to generate a state of conviction so that we can enjoy each activity in a coordinated way with the members of my team and other teams without risk of losing enthusiasm. Once we have discovered that level of passion for creating alternative solutions to the problems we face, we will have come to understand what is behind the entrepreneurial culture: a philosophy of life.

In an effort to contribute with a different idea to those already mentioned in the concept of systemic thinking, we took the liberty to quote another important source in the construction of highly effective work teams—as a reference to Covey's approach.

In the book *Power of Two* by Rodd Wagner & Gale Muller, they present eight key ideas of how people should function from a holistic or systemic point of view on high performance teams:

1. **Complementary Competences, or Associativity:** Individual strengths must add to the collective task; "wolves hunt as a team."
2. **A Common Mission:** That motivates the team—coinciding with the Senge's concept of vision)
3. **Justice:** Respect and ability to temper the feeling of displacement; understand the importance of building strong human bonds.
4. **Confidence:** Taking team risks involves understanding who will support you when you are not present.
5. **Acceptance:** Admit that one's truth (sometimes) is neither the truth of others nor the absolute truth.
6. **Forgiveness:** We are imperfect. Therefore, we must understand that each of us do our best to perform the tasks entrusted on us from our own perspective).
7. **Communication:** The base of all others.

8. **Altruism:** Disinterested, when your participation has no interests or selfish postures. There is an awareness of selfless and reciprocal reciprocity.

The interesting thing about this complementary proposal to the systemic thinking of Senge is that, at all times, the eight variables speak of human abilities during the integration of the teams. We often talk about *"hard skills"* and the impact on the use of technology, the development of innovation and the follow-up to the execution of processes. However, nowadays, the need to develop soft skills is more recurring in the discussions with the company managers. In general, we are witnessing a more *"humanized"* approach to labor. It is migrating from a stream of engineering and management of industrialized processes to one where the factor of innovation and human dignity has much more weight during the companies' decision making process.

The systemic approach cannot stop at the efficient coordination of processes, tasks, and administrative areas, it also involves the connection of human thoughts and feelings.

STEP 6. THE BIG QUESTION: WHAT IS YOUR PURPOSE?

The previous chapter presents the four initial steps related to the concept of ignition:

1. **The Awakening of Consciousness:** Awareness / sensitization
2. **The Source of All Inspiration:** Your motivation
3. **Exploring Opportunities:** The megatrends
4. **The Element of Success:** Your attitude.

In this chapter, the next step presents a new challenge:

5. **The Importance of Building Together:** The Guide Team

This step has to do with a common purpose (it can be vision or mission in some cases) for the project in which we are working (be it personal, family, work, or community related); Simon Sinek, 2017 described it as the *why* of an organization.

We want to start by emphasizing this approach in Figure 8.

Figure 8.
Source: Portillo, E. & Moreno, T. (2016), The Stragile Conscious Vision.

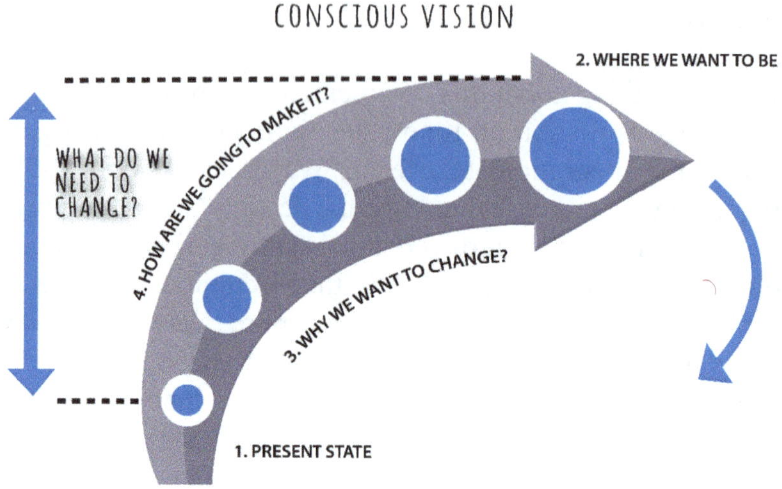

We could explain with examples the application of each one of these elements according to: personal, familiar, labor, or community-related goals. However, in an attempt to make the reading more dynamic, we decided to do it in a generic way: in the understanding that each person would have to adapt the model according to his or her case or moment in life.

As we have already mentioned, every process of transformation begins with a state of awareness that allows us to define where we are and where we want to be. Comprehending and defining the starting point will always be the foundation of our model.

If the initial variables that make up the conceptual map of our problem are not clear, most likely the path we follow will be the wrong one. It is important to remember that *"a well-defined problem is an almost solved problem"*, which implies precisely that understanding the problem we are dealing with will give us clarity when it comes to generating innovative solutions. However, at this point, it is good to remember another phrase *"much analysis, produces paralysis."* Many of us dreamers, engage endless philosophical dialogues with ourselves. The creative opportunity that our imagination gives us often distracts us from the pragmatic reality of execution. Imagining and building go hand in hand.

We would like to begin this chapter, matching the vision of two great personalities Neil Armstrong and Nicolás Machiavelli. As bizarre as the idea may seem, we believe that in the end it may make sense to explain our point.

Men judge generally more by the eye than by the hand, for everyone can see and few can feel. Everyone sees what you appear to be, few really know what you are. (Machiavelli, n.d.)

It suddenly struck me that that tiny pea, pretty and blue, was the Earth. I put up my thumb and shut one eye, and my thumb blotted out the planet Earth. I didn't feel like a giant. I felt very, very small. (Armstrong, N., 1961)

In the end, our eyes only allow us to see what our paradigms reach to understand. The key is to close our eyes so we do not affect our sight and daydream to reach other levels of idea.

The next key element of these quotes invites us to idealize our reach. The challenge at this point is to have the ability to see beyond what our eyes can see. Seeing the sculpture on the potential of raw marble is the first step to create a masterpiece. Thinking big will probably lead us to reach big. If your goal is to live on Mars, surely your first step is to build your own rocket.

During multiple courses and seminars that we have taught in Mexico and Latin America, a trigger question that never fails is: *What is your purpose in life?* Most of our students share spontaneous answers that commonly lack vision. Of course, the objective is not to expose them, but to detonate a deeper conversation.

Therefore, when their answers are "*to become an engineer, find happiness, to form a family*", or any other answer that may sound generic, we always repeat the phrase:

In life you have to look for happiness, and happiness means giving content to life. (Jose Mujica, n.d.)

Finding our passion, is what helps us get up each day with the best possible attitude and will always be the spark that ignites any bonfire. Whether this is an idea, a project, a book, a proposal, a company, a community, or love (from the broader definition of the concept, not from the sickly definition of chick flicks). Without forgetting of course that life is composed of moments and that every moment offers us the opportunity to reinvent ourselves every day.

The vast majority of students usually give another type of answer to that question: "*I do not know.* "

This answer is not necessarily bad. To some degree, it is the answer we expect, and the opportunity (from a teacher's point of view) to ask reappears. If we do not know where we are going, is the same if we move from left to the right, to the front or to the back, then the key question is: *Why do you keep moving if you have already arrived?*

To go through life like a piece of paper across the wind, with no direction, is perhaps sadder than having small goals. At least in the latter you have a limited level of awareness as to where to go, or you may pass thru life like another mortal. If you do not know what your purpose is then, *who is supposed to define it for you?* If you do not know then, *how are you supposed to find happiness?* Is it not the greatest human misfortune? To have the wonderful freedom to choose, however small or incorrect that your goals could be, is better than to throw away this great gift and to go through life waiting to see what happens.

It comes to my mind another great personal experience. Just when I was studying my professional career and when my goals and objectives lost focus due to personal issues (I now recognize that I was standing in the best opportunity that life offered me and I lacked the capacity to understand it), my father shared with me an idea:

The port of mediocrity is reached by the ocean of constant comfort. Please do not sail through those waters. (Jesus Portillo, n.d.)

At that moment, as I said I couldn't recognize the great opportunity before me, and it is not a feeling of regret that invades me now. Perhaps, if I had not lived what I had to live, today I would not be sharing these ideas. It may be, though, that the only thing I regret is that almost 30 years have passed since that time, and then I did not have the level of knowledge and awareness that I have now, a situation of course known as "*Life*" and in which case it is impossible to change, for which I now appreciate the opportunity to share it.

However, not everything is lost. Today I am delighted to be able to share it with my daughters, with my wife, with my co-workers: Andrea, Tania, Carlos, Elias, Jazmín, Rosy, Irvin, Grace, Izcali (the Jr. Enterprise at Ciudad Juárez), my family, my friends, my students, and with all the people to which I serve as a mentor and whom I can tell to their faces "the pleasure of seeing them grow fills the gap of those 30 years."

The third element raises a question that in a one way or another we have already approached in the previous chapter: *Why do we want to change?*

At this time, one of the favorite contemporary thinkers of many of us as administrators and witnesses to the quantum explosion of entrepreneurial culture is Simon Sinek. His most important contribution, as of today, is known as "*The Golden Circle.*" The reflection he shares in the third most-seen video in the history of Youtube, clearly exposes the need to define the why (Sinek, 2017).

We were more excited about calling it "*the company's DNA.*" The impact that technology has had on our lives is impressive and it would be difficult to describe it in this chapter, given the complexity of precisely the phenomenon. *Where would we begin and where should we end?* However, something that this era of open information leaves in the millennial generation (and in every generation that has been touched by this process of human transformation), is the need to collect as much data/information as possible and then have the ability to synthesize it into an idea that does not cover more than 5 words. The ability to synthesize is an indispensable skill to have today.

If we did a study and randomly took a significant and representative sample of the workforce, we would surely find that a very small percentage knows or remembers by heart the mission of the company in which they work, and the same can be said for the concept of vision.

As with strategic planning, criticism or questioning of the company's mission and vision should not be misunderstood as eliminating them from our vocabulary, on the contrary, we know that they are very powerful tools that have their specific place in business management theory. The observation goes, on the other hand, in a single idea. We must answer the following: *Why do we want to do what we want to do? What motivates us to wake up every day believing that we can somehow help transform the world? Why do we share a motivation like this with an organization?* along with a sea of people determined to move along the same path.

In some of the advisory projects with which we have worked applying the Stragile Model, the entrepreneurs have shared on several occasions a reflection "*I understood, it is like our creed!*", while opening their eyes intensely as if a great light just lit their way. It is there when we realized that the initial goal has been met: they have found their passion, their *why*. Now, comes the task to convince others to accompany them on the road, transforming this passion into reality, and do the impossible to make it transcend to the acceptance of the market reaching the sale.

We adapted the tool shown in Figure 9 according to the concept of Sinek.

The fourth step indicates a reflection that will depend highly on the life cycle in which our company (or project) is at.

If we want to transform something into a project that is at the growth stage, what we probably need to change are precepts from proposals that add value to activities that we regularly find uncomfortable doing.

Figure 9.
Source: Portillo, E. & Martínez, A. (2015), Adapted from: Sinek, S. "Golden Circle".

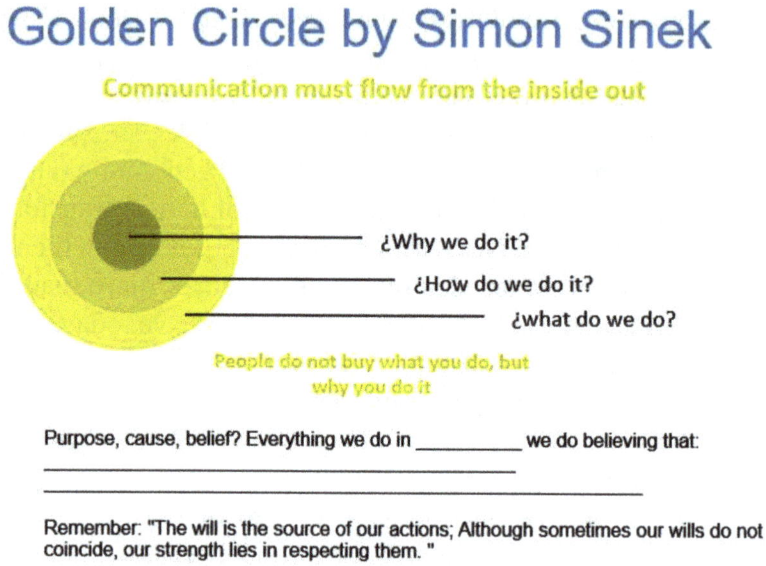

On the other hand, if we want to innovate in organizations that are in a stage of maturity, the focus will, not only be on what products or services we can improve, but in which processes of the organization we must adequate to achieve the transformation goals that we want to achieve.

For this, after completing the stage of *situational analysis*, it is strictly necessary to make a synthetic summary of the information (one of the most useful tools to do so is the SWOT analysis). After that, to conclude in an even more resumed way, what would be the eight detonating strengths of the growth of our project, in some cases known as the *Key Issues*.

Nevertheless, these processes will be explained more clearly below. It is now necessary to return to the concept of DNA.

In the article *Building You Company's Vision* written in 1996 by James Collins and Jerry Porras in the Harvard Business Review, the authors present a simple schema to understand but complicated to define in practice (Figure 10).

Collins and Porras emphasize the need to define what they call "Core Ideology" and suggest the following:

Core ideology defines the enduring character of an organization – a consistent identity that transcends product or market life cycles, technological breakthroughs, management fads, and individual leaders. In fact, the most

Figure 10.
Source: Moreno, T. (2017) Adaptation from: Collins, James C., Porras, Jerry I. (1996).

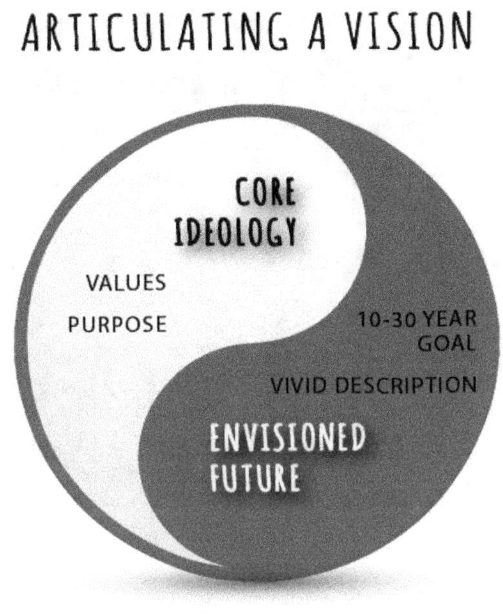

lasting and significant contribution of those who build visionary companies is the core ideology. Collins & Porras, 1996)

Even more interesting, is their proposal of what would mean the "*Core Purpose*", which comes precisely to be represented by what we define as DNA, whose most basic definition tells us that:

It is a molecule that carries the genetic instructions used in the Growth, development, functioning and reproduction of all known living organisms. (Serway et al, 2017)

The DNA is the hallmark of our vision and must run through the blood of every single one of our collaborators, so that one morning, they can experience the joy of waking up not as a punishment but as an inspiration to build a legacy for the world.

Collins and Porras also share examples that synthesize the "*Core Purpose*" of some organizations (Figure 11).

As we can see in the examples above, these are not hefty texts that no one can memorize. They are but simple phrases whose objective is not only to

Figure 11.
Source: Moreno, T. (2017) Adaptation from: Collins, James C., Porras, Jerry I. (1996).

CORE PURPOSE

HEWLETT-PACKARD	NIKE
TO MAKE TECHNICAL CONTROBUTIONS FOR THE ADVANCEMENT AND WELFARE OF HUMANITY.	TO EXPERIENCE THE EMOTION OF COMPETITION, WINNING AND CRUSHING COMPETITIORS.
MARY KAY COSMETICS	**SONY**
TO GIVE UNLIMITED OPPORTUNITY TO WOMEN.	TO EXPERIENCE THE JOY OF ADVANCING AND APPLYING TECHNOLOGY FOR THE BENEFIT OF THE PUBLIC.
WAL-MART	**DISNEY**
TO GIVE ORDINARY PEOPLE THE CHANCE TO BUY THE SAME THINGS AS RICH PEOPLE.	TO MAKE PEOPLE HAPPY.

inspire those who build them, but also serve to lay the groundwork for what is subsequently to become a unique selling proposition; that is, the source of inspiration that subsequently materializes into a solid idea of selling!

It is necessary to remember then that our capacity of vision will determine the size of the goals that we propose.

A concept that helps to clearly frame the need to take our goals to a higher level is known as Megas.

The Chamber of Commerce of Medellín offers us the following definition:

A Mega is a long-term goal, tangible and very bold, that serves as a powerful mechanism to unify efforts and stimulate progress. (Camaramedellincomco, 2015)

Again, in the strictest sense, the *Mega* should help us conceptualize our vision of opportunity. It should help us share the image of the statue behind the marble rock. So, we can concentrate on giving the first chisel!

Collins and Porras help us in the conceptualizing with some very enriching examples of this subject in Figure 12.

"The pumpkins will find their place during the trip": The first pieces of the puzzle

Figure 12.
Source: Moreno, T. (2017) Adaptation from: Collins, James C., Porras, Jerry I. (1996).

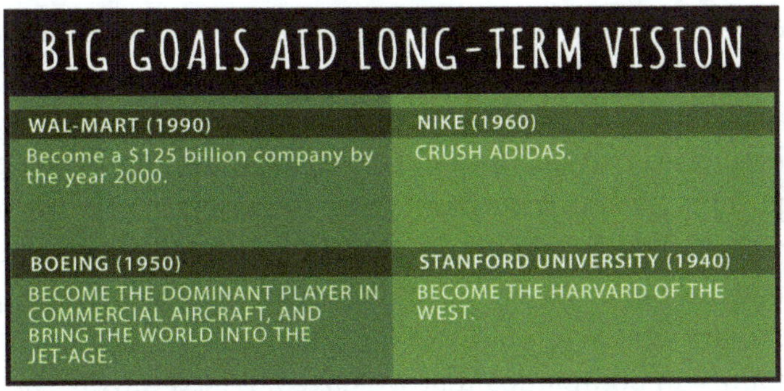

In order for the definitions and initial tools to start having practical meaning, it is necessary to start from the approach shown in Figure 13.

It is important to define the path, but it is equally important to start on it. If we stay meditating we will never yield fruits. Nevertheless, you also have to develop a strategy to walk that road! It's not about running like crazy, either. It would make no sense to run a marathon at the speed that you would run a 100-meter race; the strategy is totally different from one goal to another.

The firm ZS Associates generally shares what they consider to be the synthetic map of an organizational strategy. The first step, as we have also emphasized, is the construction of a vision that answers the question: *What do we focus on?* Subsequently, suggest the need to define an organizational structure, the inclusion of human talent and perfection in the execution of the strategy, which also answers the question: *How do we make money?*

In several of the companies with which we have worked, the three elements that correspond to the implementation part of the strategy are usually either

Figure 13.
Source: Portillo, E. & Moreno, T. (2017). Adapted from ZS Associates, 2006.

not well defined, incomplete, or do not meet the minimum efficiency requirements for operation.

There are several cases of companies in which we have found positions created for the wife, children, friends and, in the more serious cases, to maintain private personal relationships. The equation does not work like this.

The strategic vision requires a solid organizational architecture that meets the basic requirements demanded by the key issues and that also considers the gradual growth of the organization—it must begin with the structural basis of the company and later, as the resources open possibilities, begin scaling the structure to the next levels to meet the demands of the market. On several occasions, mainly in companies that we have advised during the incubation stage, we come across cases of professionals who have left their companies with the desire to set up their own business, and have the idea that in order to be successful, it is necessary to start with a complete structure of: finance, marketing, legal, human resources and production; all at the very beginning!

Logically, the problem has been that, even though they may have an interesting business idea with a lot of potential, these types of structures do not allow them to last more than 6 months and end up draining the resources of the entrepreneur during the absence of sales. Again, not because the product or service was bad, but because the market is slow to react to a proposal that is unknown.

As if this wasn't enough, an additional problem surfaces. By acting in good faith, they include family and friends in key positions. Situation that leads to confrontation because the people they selected did not have the skills that the roles they were asked to do require, or as we said before, they created positions for them, without these being necessary for the organization.

Reason why the model proposed by ZS Associates, very clearly states the process to follow in the construction of a successful company:

First, the creation of a vision; second, the definition of an organizational architecture. Once this has been defined, the third step is to look for the best human talent to fill these defined spaces. Finally, the fourth step, and also where a significant number of companies fail, is the implementation of the strategy, of which will be discussed later in more detail.

Let's remember our model and let's identify why the ZS Associates Model becomes essential (Figure 14).

Here we can see clearly how, after focus, alignment is key!

However, before moving on to the concepts of organizational architecture, the integration of human talent or the importance of execution, let us take the first step: the DNA or focus.

Figure 14.
Source: Portillo, E. (2016), The Stragile Conscious Corporate Innovation Model.

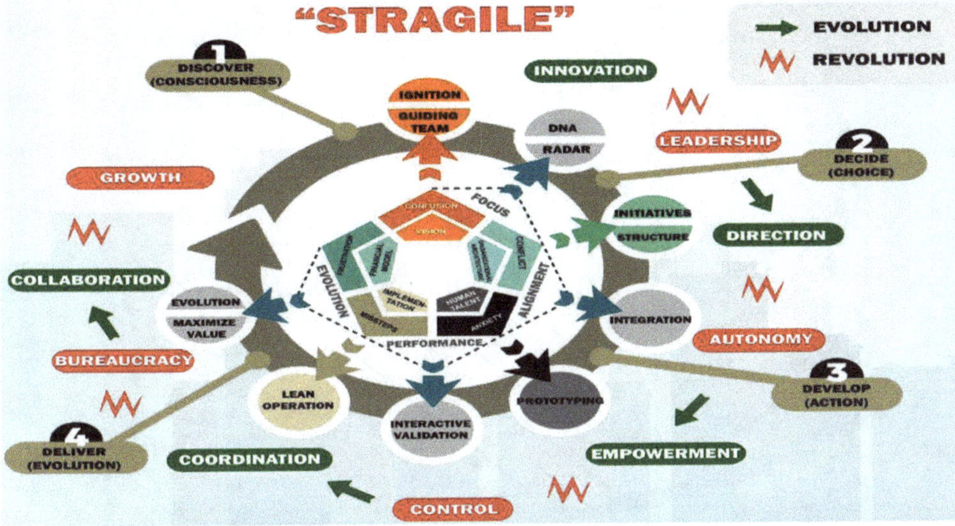

In a business forum in Mexico City, where we attended a talk by Dr. Michael Porter on the issues of competitive advantage and differentiation of products, he asked the audience the following question: *Are you competing to be unique or competing to be the best?* The murmurs among the audience did not wait. We turned to face each other as if to seek support from our neighbor, we each had a different answer but, a fear to expose ourselves publicly and to fail, suggested us to act with caution. Cleverly, Porter asked us to raise our hands if we agreed with one or another position, half of the audience were in favor of being unique and the other half were in favor of being the best—logically I raised my hand in favor of the correct answer.

After a few minutes of hearing arguments defending both positions, Porter categorically stated, "*when we talk about competitive advantage, we talk about innovation; we talk about differentiation. You should not go through life by turning back seeing where your competitor is coming. You should not lose focus. On the contrary, you must keep looking ahead, where your future is.* "

Porter talked about the importance of understanding the life cycle of the pioneer. Innovation usually gives you a gap to capitalize your ideas, when the market reacts you will begin to live the consequences of an open economy; which is not bad at all. It can help in two possible ways: to strengthen your value proposition through experience or, to explore new paths of innovation.

Figure 15.
Source: Portillo, E. & Moreno, T (2010), The Pioneer Lifecycle.

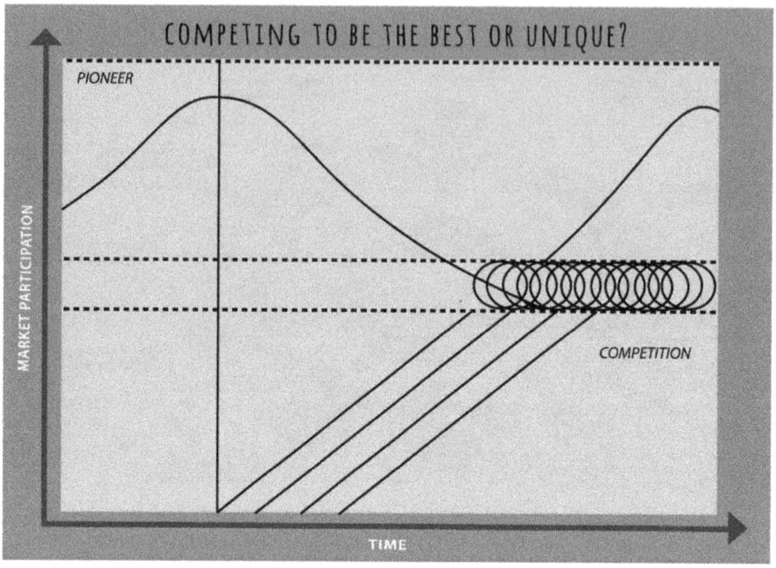

As a continuous marketing scholar, a basic question came to my mind: *Which of the four P's of the marketing mix is the one that triggers innovation?* From the answer to that question, I began to seriously question the traditional marketing model that tells us to develop practically each and every one of the 4P's at the same time—including the one I have added since then: the "S" for service.

About 10 years ago I realized the similarity that existed between marketing and the new currents of understanding that were beginning to appear.

During the process, I found another important piece of art *The Discipline of Market Leaders* by Michael Treacey and Fred Wiersema (1997).

In one way or another, the authors present an idea of how the attractiveness of our marketing offer should focus on the quality factors or characteristics that mark our differentiation in the market. In other words, in order to distinguish from the rest, they suggest to define clearly the attributes that the market seeks and appreciates in our products, the characteristics that make our services truly memorable and, above all attractive. Finally, the perspective of exchange between a supplier of products and services and the end customer. From this proposal, we can understand that, as long as there is no differentiation of value in terms of product and service, the market will tend to look for the third key element of the proposal: the price.

If this is what determines the conditions of participation of a product in a given market, then the risk of starting a price war between competitors will sooner or later diminish the aspirations of some of them, consolidating in the long-term position of one or two.

The previous reflections seem to be shaping the process of strategic adaptation in terms of marketing. Focus should be centered on the beginning of product strategy!

To corroborate this, we share another key idea that underlies this proposal. Seth Godin, in his book *All Marketers are Liars* (2005), explains the following idea:

There are two valuable rules in marketing strategy:

1. *Give me a product worth talking about (that creates value for the consumer).*
2. *Generate stories that the consumer wants to hear (if your company disappears, what would the world be missing?)*

Figure 16.
Source: Moreno, T. & Martínez, A & Portillo, E. (2016). Adapted from: Tracey, Michael, Wiersema, Fred (1996). The Discipline of Market Leaders.

The conclusion is extremely simple under this strategic approach perspective: Do not offer anything, if you have nothing of value to offer!

By collaborating in the writing of a chapter in *Information Communication Technologies and Globalization of Retailing Applications*, a book edited by Dr. Rajagopal (2009), he suggested, among other things, the need to begin to understand the dynamic of strategic adequacy that prioritizes both the resource and the strategy itself.

In Figure 16, we express the relationship between the levels of satisfaction and loyalty that are generated between customers and companies. The less satisfaction the less loyalty, the greater satisfaction the greater loyalty. We then expose the need to comprehend the linear strategic approach to market a good or service so that, in order to succeed launching an innovative proposal, in addition to have a product worth talking about, it is necessary to do an important emphasis on communication strategy at its launch, as noted in Figure 17. In this first phase, communication attracts.

Once we have managed to capture the attention of our target market, the second fire test will be to win market after they've tried our products. In this case, there are two possibilities: (1) our products are a success or (2) our products are a complete failure. In the first scenario, we will have acquired market with the quality of our products, in the second scenario, there are

Figure 17.
Source: Portillo, E. & Moreno, T. (2005). Self-Service Technologies, Doctoral Thesis.

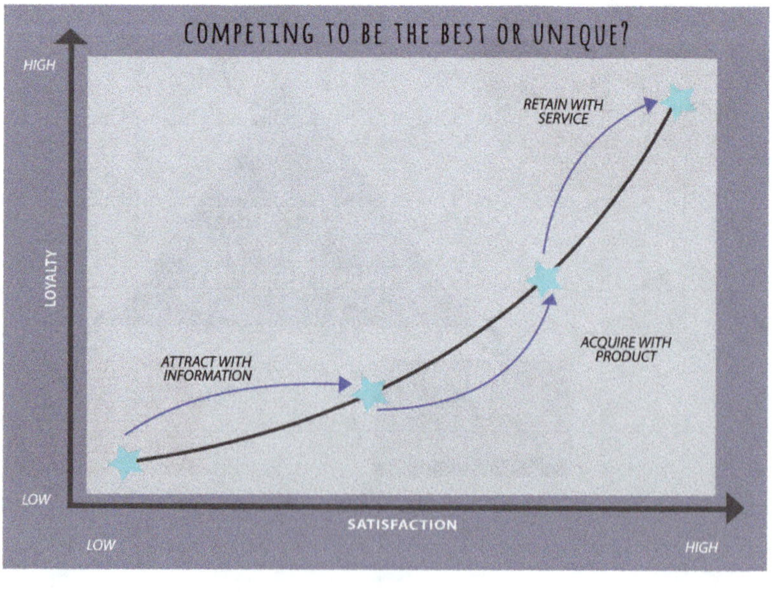

two options, (1) leave the market or (2) understand that learning promotes evolution and we give the opportunity for our idea to grow gradually.

When the product finally succeeds, the third tool will come into play to capitalize the relationship with our customers. The quality in the service is what will allow us to retain our market.

STEP 7. FROM VISION TO ACTION: NAPKINS AND PRETENDOTYPES

Recapitulating the process:

- **Chapter 1:**
    - Step 1. The Conscious Awakening: *Awareness*
    - Step 2. The Source of all Inspiration: *Your motivation*
    - Step 3. Exploring Opportunities: *The megatrends*
    - Step 4. The Element of Success: *Your attitude*
- **Chapter 2:**
    - Step 5. The Importance of Building Together: *The guidance team*
    - Step 6. The Importance of Defining a Purpose: *The DNA*

Let us first understand a structure (Figure 18) where we summarize different processes, proposals, and tools for the proposed trigger of innovation, whether these are startups (of new companies' incubation) or spinoffs (of corporate innovation).

If we visualize the process and proposals of each author and compare them with additional models related to marketing and entrepreneurship methodology (Figure 19), we will be able to comprehend how to use the proposed tools (considering the life cycle of companies) according to the need of the stage in which they are at an innovation project.

The concept is simple to explain if we understand that a fundamental part of the first phase of the innovation cycles proposed in the first chapter (adjustment problem-solution), is, precisely, market research (or Competitive Intelligence). We understand and compare the traditional marketing model against two of the most interesting approaches related to the conceptualization of ideas: the model of Steve Blank and Bob Dorf and the model of Ash Mauraya (Figure 20).

Figure 18.
Source: Portillo, E. & Moreno, T. (2016), Lean innovation processes.

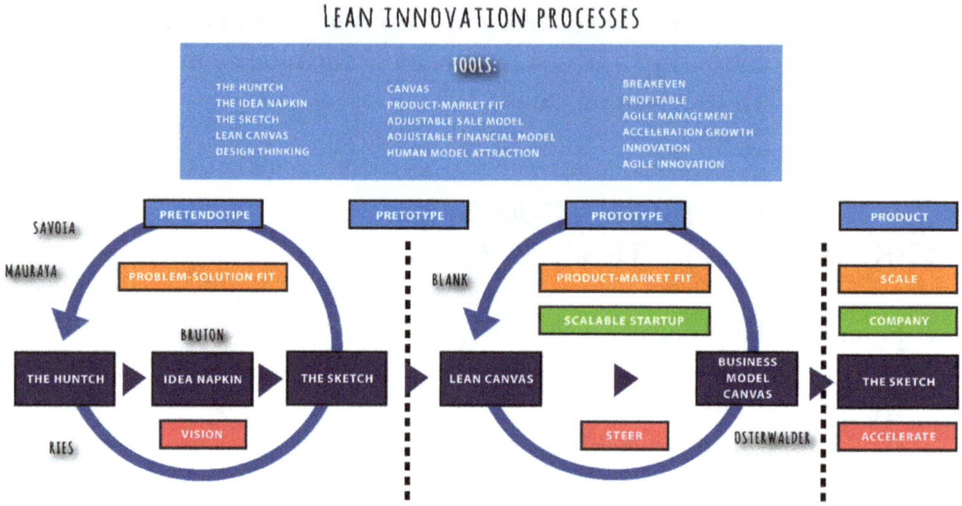

Figure 19.
Source: Portillo, E. & Moreno, T. (2016).

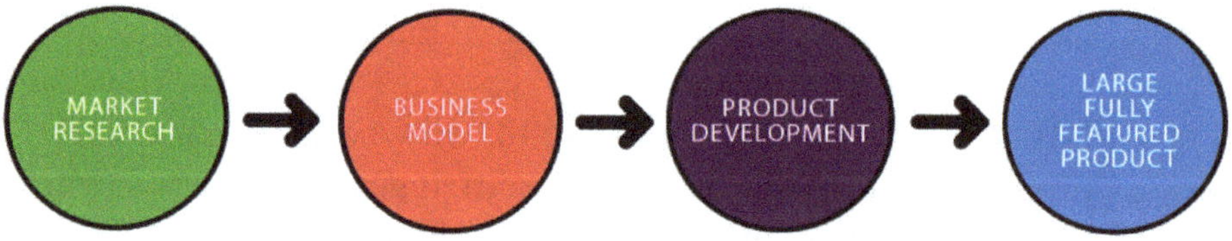

In the traditional marketing model, the proposed idea is to develop market research that allows the understanding of the buying behavior of a specific market, so that (1) its behavior can be better segmented, (2) the target market can be defined (acording to Everett M. Rogers: *Diffusion of Innovations, 1983*), and finally (3) define the positioning strategy for the brand or product to be marketed.

This leads us to understand the reason why, in the current models of entrepreneurship, market research concepts must be clearly separated into what we know as qualitative and quantitative market research (Figure 22).

The exploratory market research has great coincidences and generates the appropriate approach to explain the first phase of the innovation cycle: problem-solution adjustment. Qualitative types and techniques of research, such as in-depth interviews, group sessions, ethnographic research or tools

Figure 20.
Source: Portillo, E. & Moreno, T.(2016). Adapted from: Marketing model versus lean innovation models.

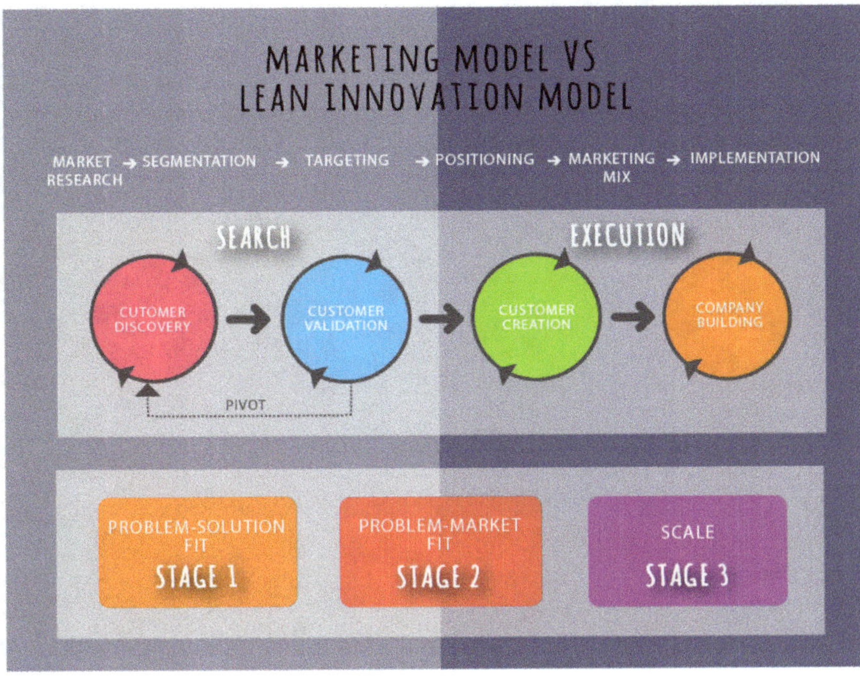

Figure 21.
Source: Portillo, E. & Moreno, T.(2016). Adapted from: Everett M. Rogers (1983), Diffusion of Innovations.

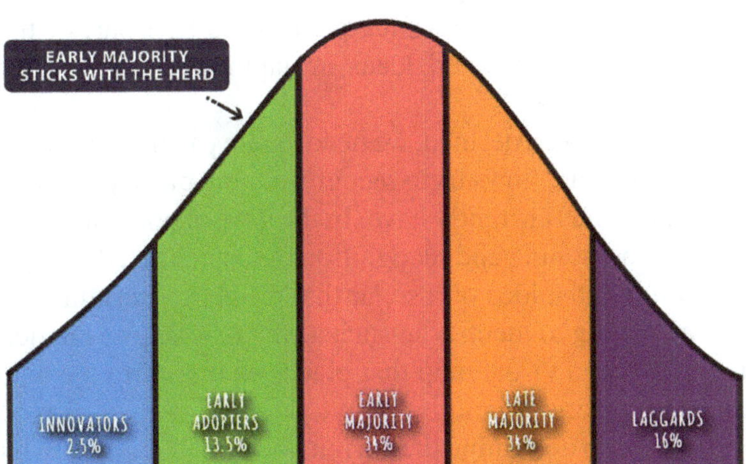

Figure 22.
Source: Portillo, E. (2016), Stragile base theories.

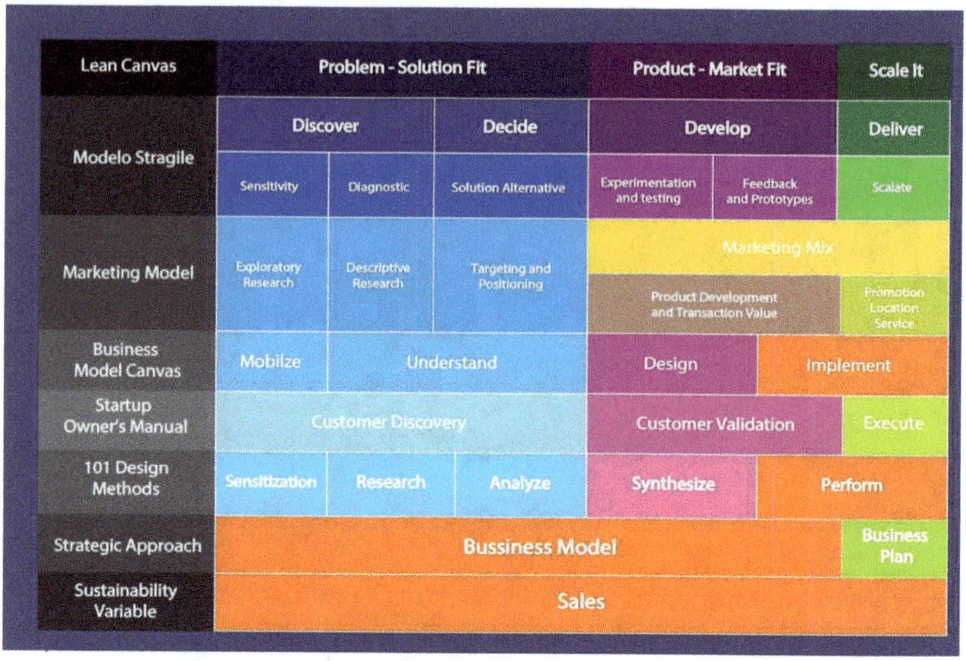

such as empathy maps, or customer journey canvases add a lot of value to the understanding of the buying behavior consumers express in this first stage. Therefore, we will briefly explain some of these tools in this chapter.

This is how all the pieces are integrated in the first part of the puzzle, we move from the diagrams and ideas in napkins (*Pretendotypes*) to the construction of *Pretotypes*.

Once the DNA has been defined, we need to move to the stage of collecting market information that allows us to recognize more clearly the position that our products and brands currently have. In the case of companies and brands with a degree of positioning and recognition in the market, the "*radars*" will work precisely under the idea of a sonar that feeds us a certain image of the surfaces that are trying to identify in such a way that helps us understand, in a summarized image, a visual map that places us in a certain context.

It is the same for organizations. Brand-positioning maps are one of the most powerful tools to determine the current status of a product in the market compared against its competition.

For initiatives in the incubation process, the tools for the construction of "radars" are totally different. Here, the use of the *customer journey canvas*,

Figure 23.
Source: Portillo, E. (2016), The Stragile Conscious Corporate Innovation Model.

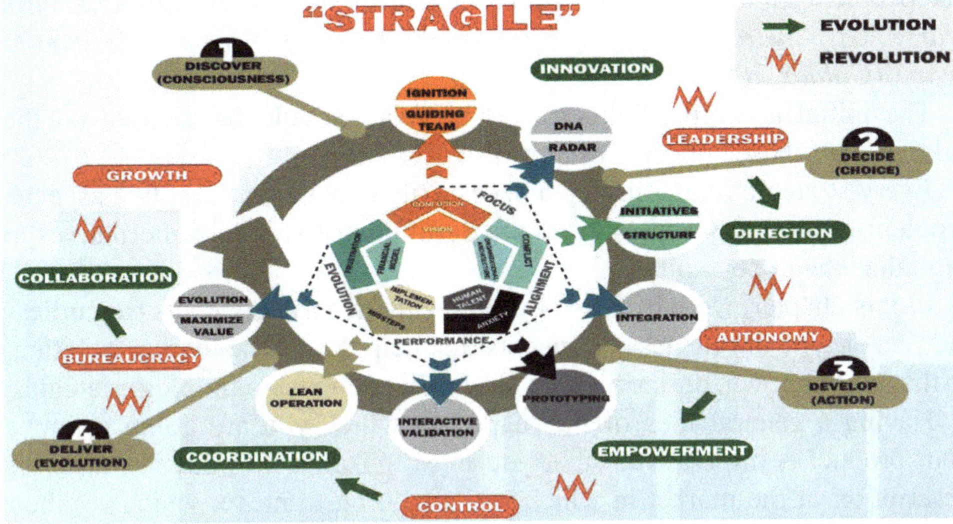

empathy maps, or any other type of current tool used in the *entrepreneurship canvas*, *innovation*, *design thinking* etc., will be of great help to map the purchase drivers of the potential consumers of a product in gestation.

Let us start by defining some necessary concepts related to the traditional methodology of market research.

The Market Analysis and the Definition of the Target Market

The Market Analysis

Within the approach of agile strategic adaptation, market analysis becomes a tool of opportunity, rather than depth.

The big difference with a model focused on lean entrepreneurship is that, until recently, *deep market studies* were of little use and high cost when it comes to developing concepts that people are not able to imagine.

Far from requiring such exhaustive quantitative studies or endless planning sessions, where it is intended to "*control*" each one of the variables that affect or can possibly affect the performance of a product in a given market, what is required are validation tools of concepts that yield immediate results regarding what consumers understand or interpret of our proposals.

At this stage, the variables product and price of the marketing mix are those that have the priority focus without neglecting, of course, the stories that people want to hear, as Seth Godin warns in his book *All Marketers are Liars*, *"Give me a product worth talking about and stories people want to hear." (Goldin, 2005)*

The initial analysis of the market behavior should be focused on the validation of these concepts as a priority.

Every strategic marketing plan necessarily contains at least two aspects: problem-opportunity analysis and the definition of strategic alternatives to capitalize the opportunity.

In this chapter, we will begin by analyzing generic concepts (regardless the size or type of business), concepts that help define the market situation, so that, once known, define high and immediate impact marketing strategies.

Having a general idea of the market in which you are going to move your product is imperative. First, you must make an introduction with the description of the market in which you are going to move, which are their principal characteristics, if there are entry barriers and what risks do they pose.

It is also important to take into account the legal, economic, social, and technological conditions that may affect the start-up of a business.

A useful tool for analysis and synthesis of information is shared by William Stanton in his book *Fundamentals of Marketing* (2007) (Figure 24).

Figure 24.
Source: Chacon, J. E. (2016). Adapted from: Stanton, W. (2007) Fundamentals of marketing.

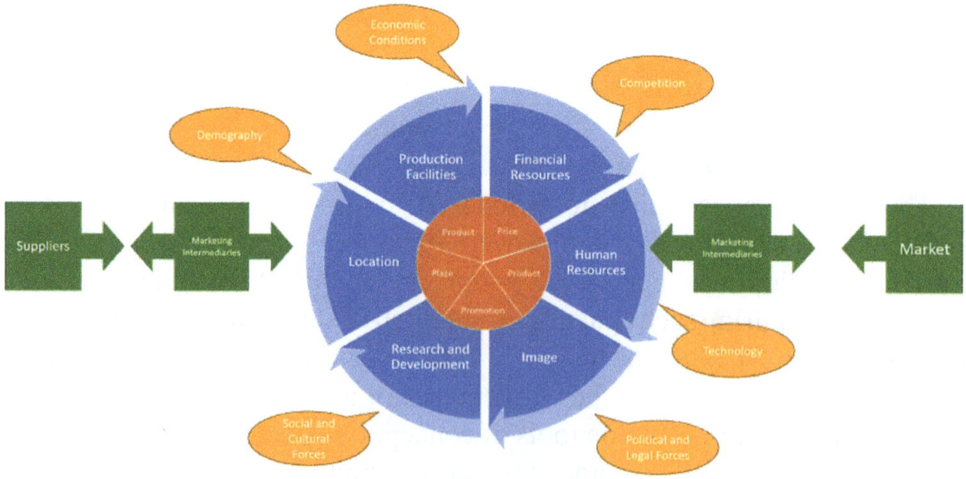

In a simplified way, Stanton develops this model in which three generic components of study can be visualized:

1. Forces of the External Macro environment to the Organization
2. Forces of the External Micro-environment to the Organization
3. Forces of the Internal Environment of the Organization

This is an important first step to begin to decompose and comprehend the problem you want to address. In our experience, we have witnessed that most people believe they know every one of the elements that make up the problem or opportunity their companies are facing.

We commonly face reactions like "*Ah yes, I already knew.*" The inevitable question in our behalf is: *Then, why do you need us?*

Alexander Osterwlader, in his book *The Business Model Canvas* (2009) presented an idea that clearly frames this situation (Figure 25).

When problem ambiguity becomes clear and turns into an opportunity thanks to knowledge, research, and focused human talent understanding the situation; the immediate reaction is: *"I already knew that!"*

We have interpreted this expression in two ways:

1. **Positive:** *"It's true, it's so clear that although I live it every day, I did not realize the answer I was looking was in front of me the whole time."*
2. **Dangerous:** "*Yes, I know. Tell me something I did not know before.*"

At first, people generally appreciate your guidance as a consultant, given that an external approach helps them clarify the vision before their

Figure 25.
Source: Portillo, E. & Moreno, T.(2017). Adapted from: Osterwalder, A. & Yves, P. (2009). Business Model Generation.

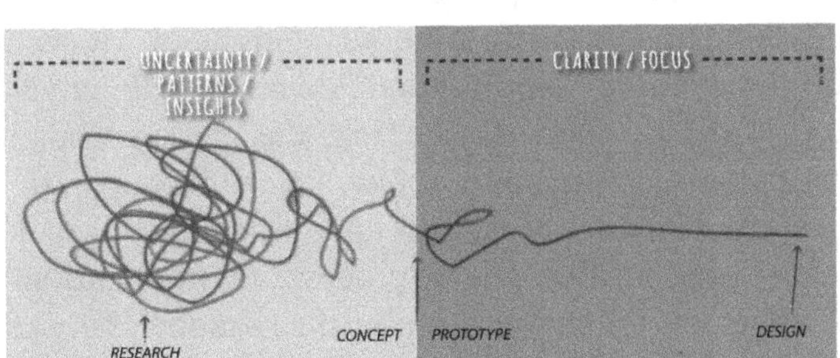

administrative myopia in which they tend to fall when the shift between strategy and operation becomes routine.

In the second case, we call it "*the dangerous way*" because people's ego, perhaps the most common expression of human fear, does not allow us to accept omissions or mistakes that we, as human beings, have the "*right*" to commit. Our educational system and our culture in general, have taught us that failure is unacceptable, which is far from the human condition. We must accept that, to a great extent, success is given by two characteristics of the human being: *perseverance* and *resilience.*

Gerald Jampolsky in *Love is letting go of fear* (2003) states, *"Our ego motivates us to believe that we will be in a more peaceful and happy situation if we find someone to blame for what is happening to us and in who we can deposit our anger. So, we develop and maintain an attitude that we think is healthy."* In this way, we develop a behavior of justification when something goes wrong, we point out that what happened is not our responsibility.

Curiously, when we see handicapped people succeed, we tend to admire them and say "*What a character he/she must have to reach that position*" when the truth is that the person had no other choice but to face the life it had, without being able to blame anyone. Confronting its fears and facing its failures in a different condition than anyone else, at the end, is what made them develop higher abilities than the *"normal"* individuals, because we do not have the need to try harder in order to achieve what they had to achieve with great effort.

Demand Analysis

The analysis of the factors or forces that influence the buying market's behavior is important. However, estimating the potential of a market is, perhaps, the most important variable in the initial stage of a company or the development of a new product.

One of the most interesting proposals regarding the market behavior is exposed by Everett Rogers, with some adaptations of Moore is shown in Figure 26 and some others from Furr and Ahlstrom in Figure 27.

From these, we must learn that at first, the market tends to attend to innovation in a reserved way. Only a few are willing to "*risk*" before the opportunity involved in innovation. This approach allows us to interpret how the demand should be considered in the process of introducing an innovation, as well as the effect this will have on our business model. Furr and Ahlstrom

Figure 26.
Moreno, T.(2017). Adapted from: Moore, G. (1991) Crossing the Chasm.

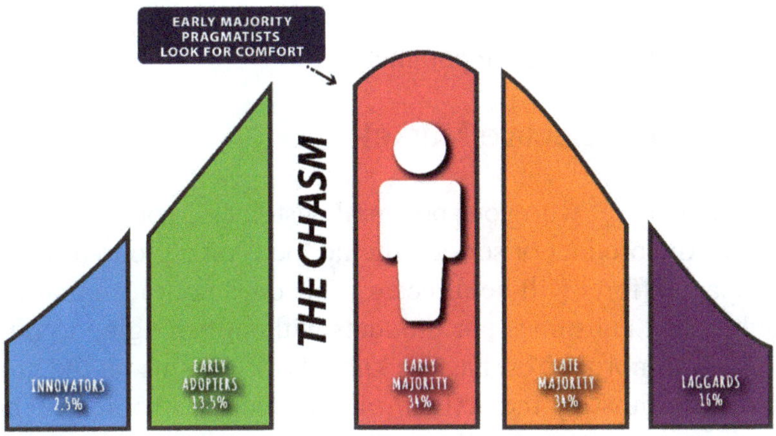

Figure 27.
Source: Moreno, T.(2017). Adapted from: Furr & Nathan & Ahlstrom & Paul (2009). Nail it Then Scale it: The Entrepreneur's Guide to Creating and Managing Breakthrough Innovation.

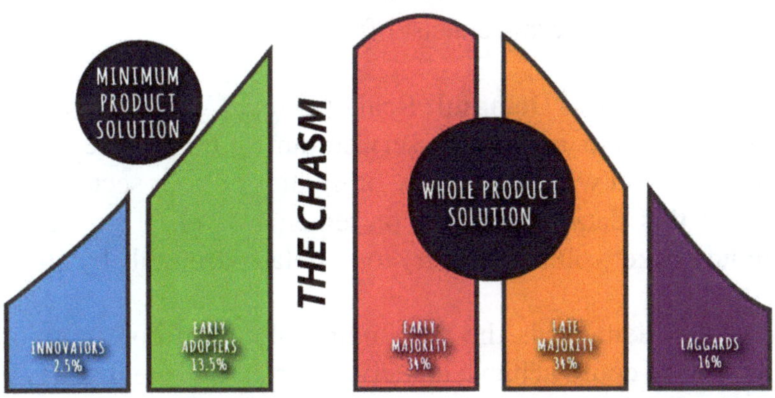

were even clearer by (1) identifying the need to develop a minimum product solution on a smaller scale, (2) validating it (as several authors like Blank and Dorf have agreed on), (3) pass through the abyss of innovation and, only then, (4) to scale.

This make us think: *What's the demand for your product or service? The growth forecasts for the sector? If it is an expanding sector, what market share do you expect to have?* And above everything you do, do not forget to support all this by means of references to studies or statistics that endorse your forecasts.

It is imperative to gather as much knowledge as possible of current, potential, and future demand. By studying the state and evolution of these parameters we can have very valuable information to adapt our business to the state and evolution of these parameters.

Definition of the Objective Market

You must ask yourself who your potential customers, your target, and your tribe will be. Any product or service should focus on a group to which most of our company efforts will be directed. This does not mean that there is a certain public that consumes our products without belonging to our target, but it will be a minority. You have to know how to focus on the group that interests you the most by the activity of our business. To define the target audience, we must answer the following questions: *Who are my clients? What are their characteristics? How many are they? How have they survived before having our product or service? Why are they going to buy from us? Which segments are the most attractive for our product?* To define our audience, we can use information of the real and potential demand.

Let's see what these terms consist of:

- **Real and Potential Demand:** Real or "*actual*" demand is the demand that exists for our product or service during this or the last year. To calculate it, you will have to rely on studies of your sector that usually calculate this data at national or regional level. If you do not find accurate market volume data, try to calculate it roughly by other means.

The potential demand is the maximum volume of consumption that a product or service could achieve in a given period of time. It parts from the idea that all potential buyers of the product would become buyers with a certain level of consumption. In fact, it is a theoretical concept that is used to compare between different types of demand.

For example, let's think of the case of a dentistry clinic. The real demand is the number of people who come to the clinic during a year, being the potential demand all the people that could go by necessity to the clinic.

- **Future Demand:** Is the expected demand in the following year and involves the task of calculating the percentage increase or decrease in relation to the current or actual demand.

It is a very important data for the planning of companies, which usually calculate for periods of three to five years, and that helps them in their strategic plans.

If it is estimated that future demand will increase by 20%, we must adapt our business to a possible increase of this size.

Viewed graphically, the best approximation of a strategic approach to the target market (niche, target market, potential market) would be as shown in Figure 28.

- **Market Participation Goals:** In this section, we will try to calculate the market participation goals based on the data obtained regarding the *actual*, *potential*, and *future demand*. It is about calculating how much share of the market we can get in a given time.

As in the other sections of the business plan, you must be realistic and set a viable goal. It is absurd to think that in your first year and without data to support your theory, you will have a market share of 50%, as expecting half of all people who buy that product in the country will buy it form you.

Figure 28.
Source: Portillo, E. (2014), The Stragile Business Model.

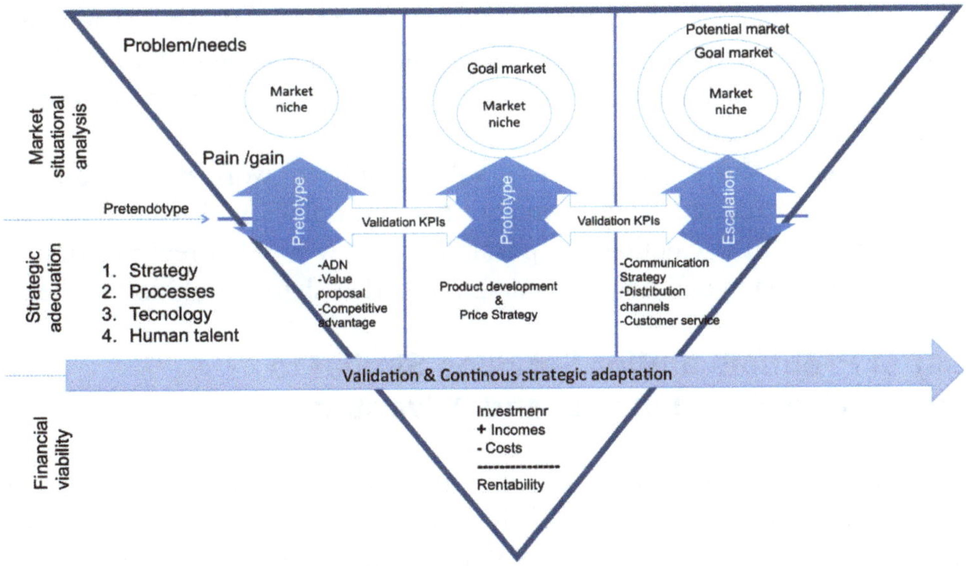

- **Competition Analysis:** This point is crucial and the more you develop it the more aware you will be of what you are going to face. In addition, it shows the person reading your business plan that you know the competition to a grear extent and that you are well aware of where you are standing.

Unless your product is unique in the market, an invention that covers a need that did not exist until you arrive, you must study competitors in detail.

- **Competitor's Data:** In a way, there is always competition. Since, even if you are offering something new and innovative, the needs you meet will force you to ask yourself the following: *Who is my competition? How is it? What are my strengths and weaknesses? What plans does my competition have? From who am I going to take away customers?*

Market Penetration, Employees, and Geographic Location

Remember that you must first set your answers on reliable, non-speculative data. Try to support your conclusions with studies of business associations in your sector, official statistics, etc.

The more data about your competition, the more knowledge you have about them. Do not think that contributing data on your competitors will hurt you. The market is very big and there is room for everyone. The more data you give and, above all, the more solutions and strategies to position yourself in relation to them, the more credibility you will have in front of your potential investors.

Try to make a study in terms of total sales, number of workers, delegations, technological level, etc. Then, compare these variables with your proposal and try to do all the calculations possible to know in what real position is your competition in and how you can gain market share in relation to them.

List of Products and/or Services Similar to Ours: Price, Deficiencies, and Advantages

In order to conduct a proper study of your competitors, you must gather all possible data. For this, the best way is to test your products and/or services. Access all the public information available to the company and all kinds of techniques that allow you to know more in order to to enable you to answer questions about who you are, how you are, how you got here, and where you are going.

Weaknesses and Strengths of Each Competitor

What you are most interested in knowing about the competition are the similarities and differences between their product and yours. Try to clearly explain the shortcomings and advantages compared to your product and the price difference between them. It's about showing that you're not going to do exactly the same thing as your competitor, in the same way. You will compete with a product with different variables, whether they are quality, price, innovation in marketing, etc.

When two fighters face off, the first thing they do is to study each other to discover their strengths and weaknesses. The market is a continuous struggle and this study between competitors never ends. You always have to be on the lookout and not lower your guard. Learn from the virtues of your opponent and try not to make the same mistakes. Try to reflect on this point the strengths and weaknesses you find in each of your competitors.

The main factor that determines the success of a business is knowing how to make the right decision.

Identifying Key Factors in your Competitors' Success

By analyzing the keys to your competitors' success, you can intuit what may be the key to yours, and direct your efforts in that direction. It is possible that the success of your competitor derives from the fact that he either was the first, settled in a good area, innovated in distribution, or because he improved the product... study thoroughly what his keys to success were since it can help us find ours.

Comparative Analysis Between Competitors Based on Previous Information

Once you have collected and analyzed all the possible data regarding your competition, it is time to do a reflexive analysis. Compare each of the companies you've analyzed between them and your business and try to group or classify your competitors according to certain parameters. It is a matter of summarizing what was stated in the previous points and draw conclusions about it.

Additional Useful Tools in Collecting Information for the Radars Construction

Empathy Map

One of the most relevant contributions of the canvas models is the ability to synthesize. The ability to display in a filtered, valid, simplified, and executive way high-impact suitable information (understanding the executive term as: the need to have accurate, concrete and current information for immediate decision-making).

An empathy map is no exception in that matter. It is a tool that allows reflection by differentiating between explicit and conscious behavior of a consumer, as well as implicit or subconscious behavior. Tools like this help to differentiate what the consumer says, does, as well as what the consumer secretly desires but represses.

Understanding the rationality of behavior is important. Knowing the emotion behind motivation or behind an attitude is the key to generate successful marketing strategies that accompany any entrepreneurship initiative. Although it is not enough to analyze the events behind the Brexit, the Referendum for Peace in Colombia, or the elections in the US, where users commented on one thing and acted differently.

Figure 29.
Source: Moreno, T.(2017). Adapted from: Lee, D. (2015). Empathy map for Design Thinking

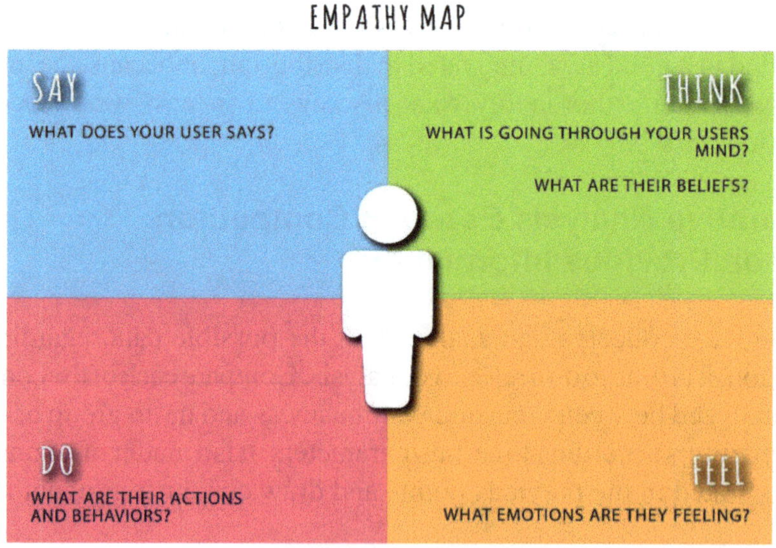

Customer Value Canvas

As in the case of the previous tool, the synthesis capability of this tool is very powerful. In this case, it tilts more for the stage of adjustment problem-solution. The reason is simple: it allows us to understand, in a simplified way, the need to know the motivations that provoke attitudes in the target market we are studying. Understanding how our market niche suffers or enjoys is highly relevant to create attributes that generate satisfaction and lessen the negatives.

Customer Journey Canvas

There are several versions of this type of canvas. It was important to create a tool that had to do with the traditional buying behavior in marketing. In fact, the basis of the model is precisely the pattern of buying behavior explained by Stanton in his marketing book. Still, we use it as an adaptation to identify which variables we should consider and study and at what time according to the decision-making process of the consumer.

Figure 30.
Source: Moreno, T.(2017). Adapted from: Osterwalder, A. (2012). The Customer Value Map.

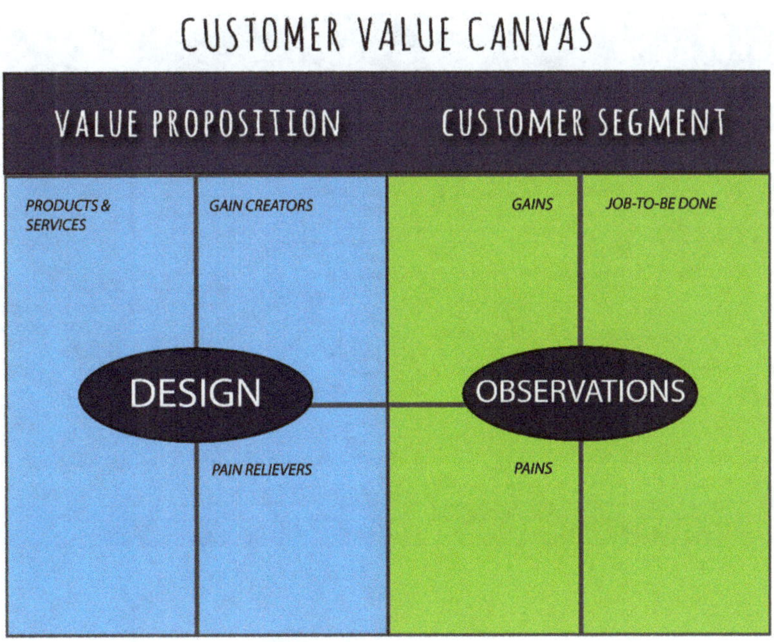

Thus, traditional marketing literature tells us that there are several factors already known (technological, social, psychological and situational) that influence the consumer motivation to purchase. There are attributes of delight and pain that either detonate or slow down the motivation, attitude, purchase intention, and purchase behavior; in that order. Finally, the comprehension of expectations and experiences as a source of satisfaction or dissatisfaction with a new product.

Fishbein Positioning Model

Fishbein's Preferences Model is perhaps one of the most useful tools in understanding the place a product or service occupies in the mind of a consumer over other competing brands in the market. While it is true that it has an area of improvement in terms of the launch of new products (since the user usually does not have a brand reference and the evaluation becomes subjective), it is also true that since it is a map of positioning (evaluating only the user's perception), when it comes to new products, it helps to evaluate the first impressions of what the user *"believes"* to be receiving.

Figure 31.
Source: Portillo, E. (2014), The Stragile Customer Journey Canvas.

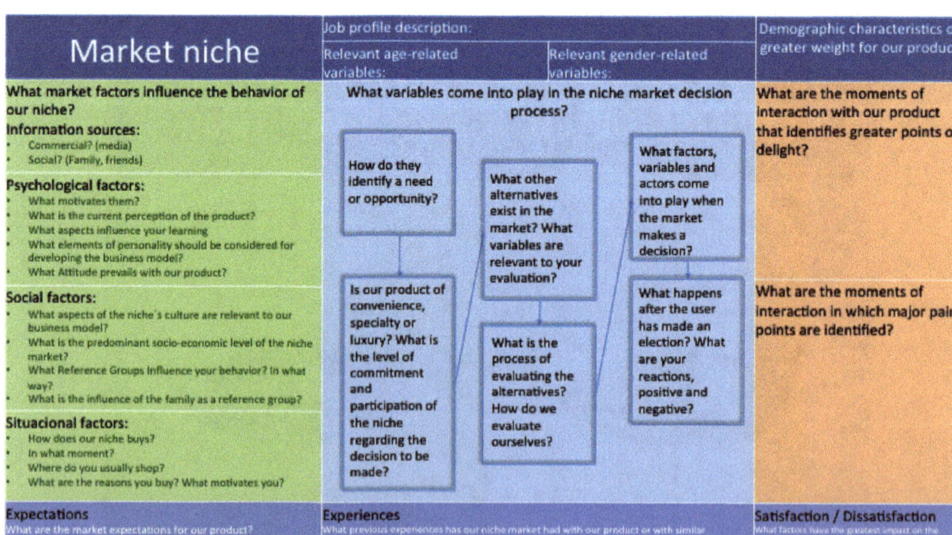

In the case of already positioned brands, it allows us to contrast how we are positioned against other brands in each of the relevant variables for the consumer.

The Fishbein Model is nothing but the sum of two factors: the relative importance of an attribute in relation to the other attributes considered important for selecting a brand and the performance evaluation of each of the competing brands.

Considerations

1. "j" can be an idea, a service, a person, etc.
2. Consumer "k" can be a segment.
3. It is convenient to analyze the three main attributes to determine the competitive advantages and disadvantages. Since it is not a rational process that we follow to make a decision of which product to buy, it is convenient to analyze separately the performance of the first three factors and its impact on the user's decision making.

Steps

1. Determine the attributes that customers take into consideration when selecting a product (group sessions).

Figure 32.

$$P_{jk} = \sum_{i=1}^{n} W_{ik} A_{ijk}$$

Where:

P_{jk} ***="k" Consumer percepction over "j" brand***

W_{ik} ***=Product´s attribute "i" importance, given by "k" consumer***

A_{ijk} ***=Quantity of attribute "i", the "k" consumer thinks that the product has.***

2. Determine the relative importance of these attributes (questionnaire).
3. Determine the evaluation for each of the attributes under consideration, that the client assigns to each of the products being evaluated (questionnaire).
4. Calculate preferences using the preference model (formula).
5. Develop a positioning map of the attributes (an example is shown in Figure 33).

Service Blueprint

In this stage of market research, having a tool like the *service blueprint* at hand is very valuable simply because it allows us to visualize where the opportunities are in the front stage as well as in the back stage of the client attention process. To be able to see the whole picture and from this summarized perspective allows us to reflect opportunely on the improvement opportunities.

Figure 33.
Source: Moreno, T. (2016), Fishbein Positioning Map Adaptation. Stragile.

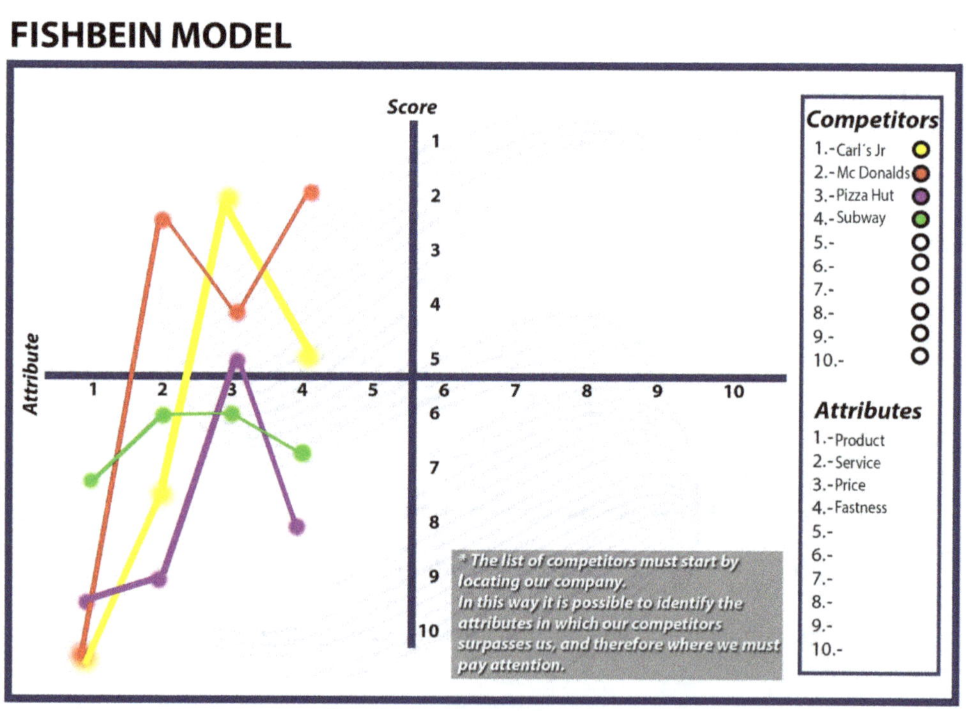

Figure 34.
Source: Moreno, T.(2017). Adapted from: Anna (2017), New Templates for Service Design, Personas and Blueprint.

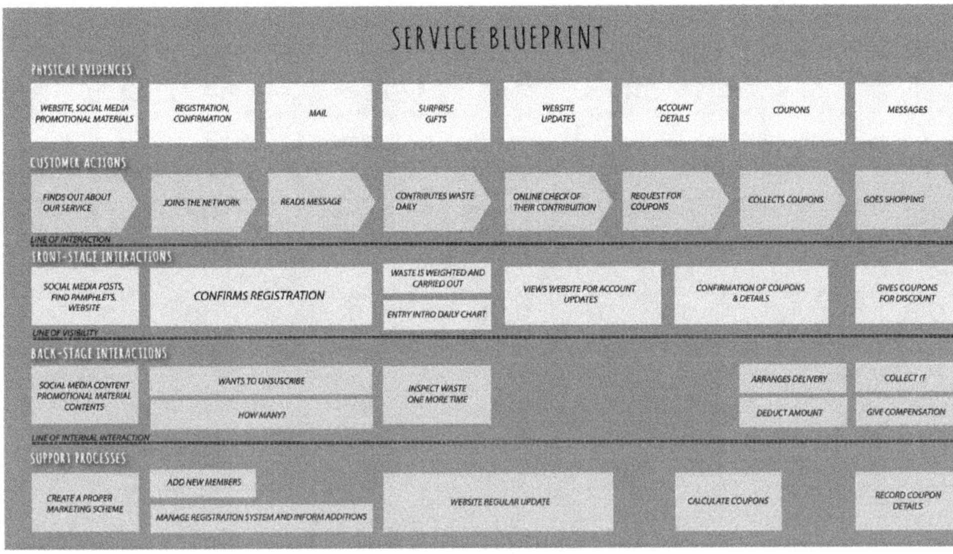

Quality Control Engineering Processes Maps

Like the *service blueprint*, a process map (with its respective analysis and quality control tool) helps us to better understand the internal production processes and the areas of opportunity within them.

Blue Ocean Strategy

Understanding the place we occupy in the market as well as if the "*ocean*" we are in is either blue or red, will give us a clear perspective of what strategies to follow in terms of innovation. As we mentioned before, *are we competing to be the best or are we competing to be unique?* Figure 36 shows another tool that can help us synthesize information (Blueoceanstrategycom, 2005).

The Einsenhower Matrix

One of the biggest challenges we have in filtering and prioritizing information is keeping our feet on the ground. Our creativity and imagination is so powerful that we could continue to imagine models and methodologies

Figure 35.
Source: Moreno, T.(2017). Adapted from: Kottler, Philip, Keller, Kevin (2012). Marketing Management. Pearson Education.

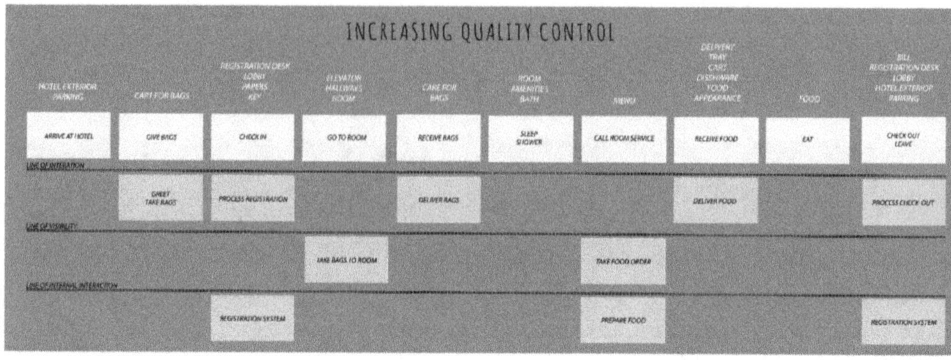

Figure 36.
Source: Moreno, T. (2016), Blue Ocean Strategy Canvas. Stragile.

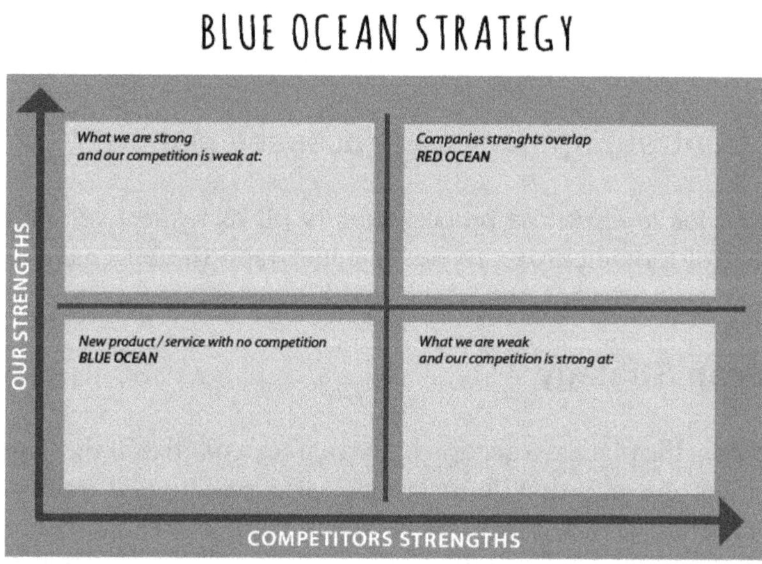

without realizing that the opportunity lies in the opportune execution, rather than in the perfection of ideas.

The Eisenhower Matrix helps us make a prioritization exercise by separating the activities that are really important and urgent and thus, defining the basis for the implementation of a plan of action orderly and especially efficient from a point of view of resources management (Eisenhower, 2011).

Figure 37.
Source: Moreno, T. (2016),Blue Ocean Strategy Canvas. Stragile.

Outside Innovation

According to Patricia B. Seybold, 2001 *"The only effective way to differentiate yourself in the market is to open up the enthusiastic consumer business and give voice to the design of every aspect of a product or service. Only companies that engage the consumer in the innovation process will be able to create products or services that better meet the potential customer, revolutionize existing business practices and models, and ensure customer loyalty. "*

In Figure 38, we share a tool that promotes an analysis of how an open collaboration platform proposed by Deloitte could be set off:

STEP 8. INFORMATION SIMPLIFICATION FROM THE SWOT MODEL: THE FINAL STEP OF THE INITIAL STAGE

Once all the information we consider relevant has been gathered, it is necessary to summarize, synthesize, and present it in such a way that, just as in a canvas, one can find in a clear way in each and every one of the opportunities that exist inside and outside of an organization.

Figure 38.
Source: Moreno, T.(2017). Adapted from: Scott, W. (2013). Rising Tide: Exploring pathways to grow in the mobile semiconductor industry.

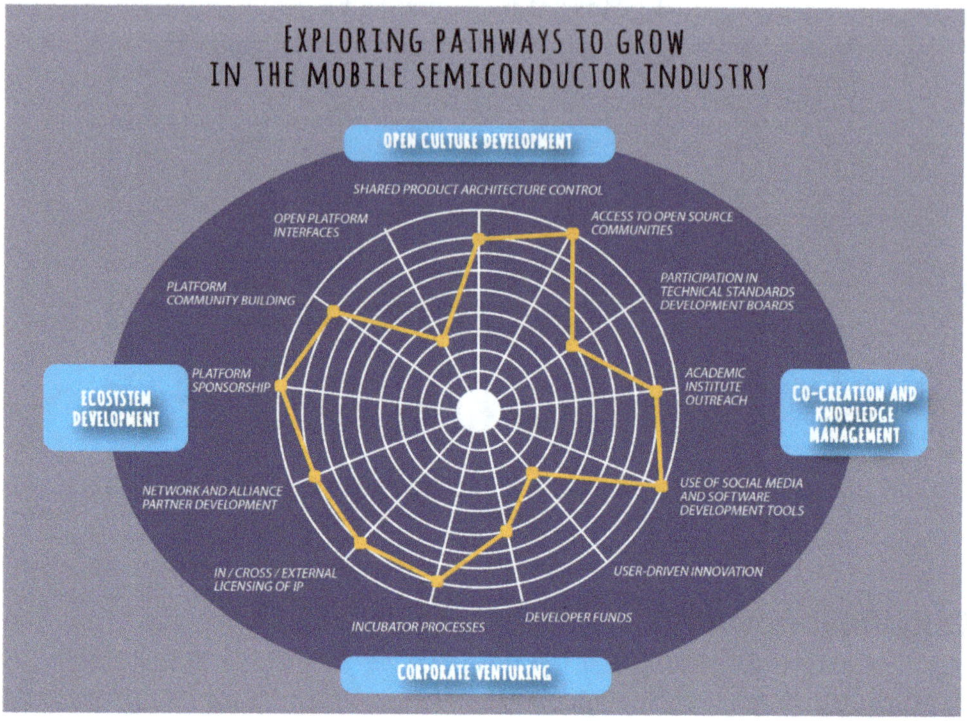

Figure 39.
Source: Moreno, T.(2017). Adapted from: Dahp-mafforg (2017).

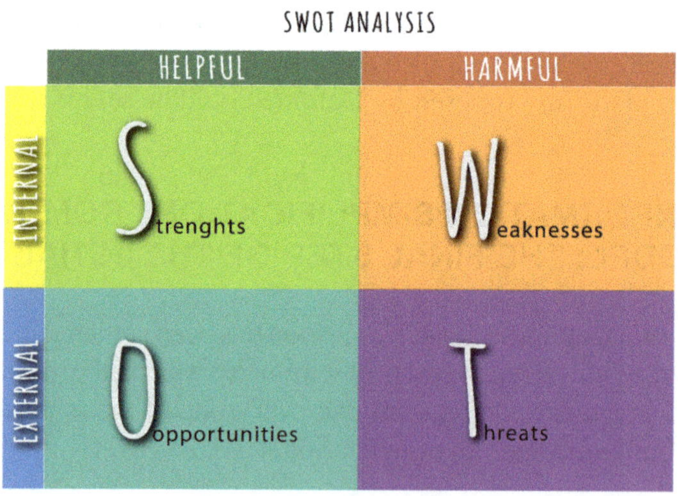

As the example in the chart below, it is necessary to separate the internal or external forces or factors. A SWOT analysis divides strengths and weaknesses as internal tools or opportunities and separates them from opportunities and threats, which it considers external forces.

The most important part of the summary of the SWOT model is that it precisely identifies the external and internal forces that have the greatest impact on the development of a business model. Forces that sooner or later will have to be converted into concrete actions or strategies to take advantage of or diminish (as the case may be), impact on our market niche.

REFERENCES

Aho, M. K., & Bennett, E. (2013). *The Machiavellian Librarian: Winning Allies, Combating Budget Cuts, and influencing Stakeholders*. Chandos Publishing.

Anna. (2017). *New Templates for Service Design, Personas and Blueprint.* Available from: https://realtimeboard.com/blog/templates-service-design-personas-service-blueprint/

Asmeorg. (2017). *Asmeorg*. Available from: https://www.asme.org/engineering-topics/articles/aerospace-defense/neil-armstrong

Blank, S. (2005). The Four Steps to the Epiphany (1st ed.). Academic Press.

Camaramedellincomco. (2015). *Camaramedellincomco.* Available from: http://www.camaramedellin.com.co/site/english/About-the-Chamber/The-Medellin-Antioquia-Chamber-of-Commerce/History/Chambers-of-Commerce-in-Colombia.aspx

Collins, J., Porras, C., & Jerry, I. (1996). Building You Company's Vision. *Harvard Business Review*.

Dahp-mafforg. (2017). *Dahp-mafforg*. Available from: http://www.dahp-maff.org/blog/swot-analysis-of-the-livestock-sub-sector-in-cambo/

David, L. (2017). *David Lee Tech.* Available from: https://davidleeedtech.wordpress.com/2015/07/08/empathy-map-for-design-thinking/

Denning, S. (2011). How Do You Change: An Organizational Culture? *Forbes*.

Eisenhower. (2011). *Eisenhower.* Available from: http://www.eisenhower.me/eisenhower-matrix/

Goldin, S. (2005). *All marketers are liars* (1st ed.). Portfolio.

Jampolsky, G. (2003). *Love is Letting Go of Fear* (1st ed.). Celestial Arts.

Moore, G (1991). *Crossing the Chasm: Marketing and Selling High-Tech Products to Mainstream Customers*. Academic Press.

Myers, T., Dorf, R., & Nelson, A. (2015). *Technology Ventures: From Idea to Enterprise* (3rd ed.). Mc Graw Hill Education.

Rajagopal. (2009). *Information Communication Technologies and Globalization of Retailing Applications*. Academic Press.

Richman, R. (2015). *Everything you know about change management is wrong*. Mayo 2017, de The Blog Revelations and Observations.

Rogers, E. M. (1983). Diffusion of innovations (1st ed.). Academic Press.

RussellB. (1933). *Russell-jcom.* Available from: http://russell-j.com/0583TS.HTM

Senge, P. (1990). *The Fifth Discipline*. The Art & Practice of the Learning Organization.

Serway, R. A. (2017). *Just the facts 101*. Chicago: Content Technologies.

Seybold, P. B. (2001). *The Customer Revolution: How to Thrive When Customers Are in Control.* Crown Pub. doi:10.1571/0609607723

Seyboldcom. (2017). *Seyboldcom.* Available from: http://www.seybold.com/patricia.htm

Sinek, S. (2017). *Startwithwhycom.* Available from: https://startwithwhy.com/

Stanton, W. J., & Futrell, C. M. (2007). *Fundamentals of Marketing*. McGraw-Hill.

Wagner, R., & Muller, G. (2009). *How to Make the Most of Your Partnerships at Work and in Life*: Gallup Press.

Wiersema, F., & Treacy, M. (1997). *The Discipline of Market Leaders: Choose Your Customers, Narrow Your Focus, Dominate Your Market*. Addison-Wesley Publishing Company.

Wilson, S. (2013). *Rising tide: Exploring pathways to growth in the mobile semiconductor industry*. Available from: https://www2.deloitte.com/insights/us/en/topics/emerging-technologies/rising-tide-exploring-pathways-to-growth-in-the-mobile-semiconductor-industry.html

Chapter 3
Innovation Initiatives

ABSTRACT

Innovation is associated with inspiration and creativity. While this affirmation is true, it is incomplete because innovation is a process that can be planned and systematically programmed and reproduced with specific methodologies, not just for the "chosen one." But we have a problem. There are dozens of innovation methodologies, all of them theoretically presume to have the "key" for innovation. Many of these methodologies give very interesting information with similar approaches, but some other propose contradictory paths. For an executive, it is simple: too much information to decide which approach would be more appropriate or where to start can be a problem, so a common behavior is to try to use the business fashion almost by faith. However, among all these amorphous and even contradictory information, if we combine three basic methodologies: design thinking (DT), quality function deployment (QFD), and inventive problem-solution theory (TRIZ: Russian acronym for Теория Решения Изобретательских Задачque), we have a powerful proven process for a successful innovation, increasing the innovation rate of success for any company. This is explored in this chapter.

INTRODUCTION

Our structure is unique: we do not operate as a large company, but as a series of small ventures, which makes everyone much more agile. Dara Khosrowshahi, CEO Expedia Inc. (Morales, 2017)

DOI: 10.4018/978-1-5225-4023-6.ch003

Let us return to the Stragile model to continue this journey of corporate innovation (Figure 1).

We have followed a linear process with the intention of providing clarity in the application of this methodology, even though we understand that the processes of generating creative solutions to problems do not follow a linear process at all (as Osterwalder described it at some point).

So far, the summary of the proposed steps has been as follows:

- **Part I Consciousness (Discover):**
    - **Chapter 1:**
        - Step 1. The Conscious Awakening: *Awareness*
        - Step 2. The Source of all Inspiration: *Your motivation*
        - Step 3. Exploring Opportunities: *The megatrends*
        - Step 4. The Element of Success: *Your attitude*

Figure 1.
Source: Portillo, E. (2016), The Stragile Conscious Corporate Innovation Model.

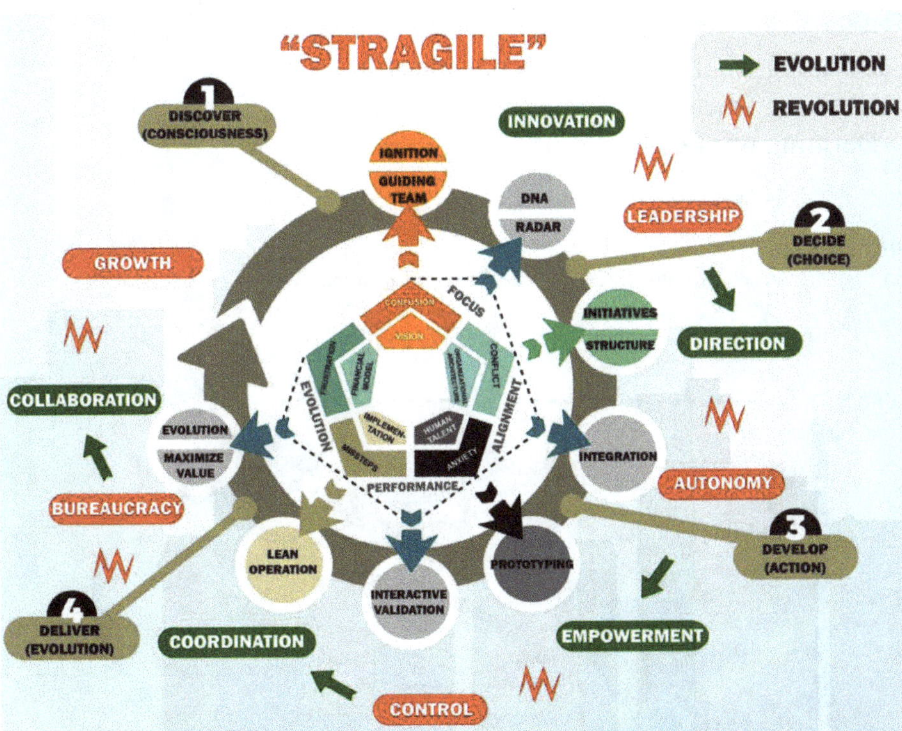

- **Chapter 2:**
  - Step 5. The Importance of Building Together: *The guidance team*
  - Step 6. The Importance of Defining a Purpose: *The DNA*
  - Step 7. Identification of Opportunities: *The ituational analysis*
  - Step 8. The Importance Behind the Simplification of Information: *The SWOT analysis*

Once the information provided by the market has been filtered and simplified (taking into consideration that the internal factors of an organization are part of the overall market dynamics), the next step will be to prioritize processes and actions that will lead us in the short term to generate results according to our initial definition of mega—or ideal situation. In the previous chapter, we focused on defining the first three relevant aspects of the Stragile approach (Figure 2).

In our adaptation of the Kurt Lewin "f*orces*" model, these first three steps are defined at the center of our chart, as shown in Figure 3.

1. Defining where we are (Current state)
2. Defining where we want to be (Ideal State)
3. Defining our DNA (Why)

Figure 2.
Source: Portillo, E. & Moreno, T. (2016), The Stragile Conscious Vision.

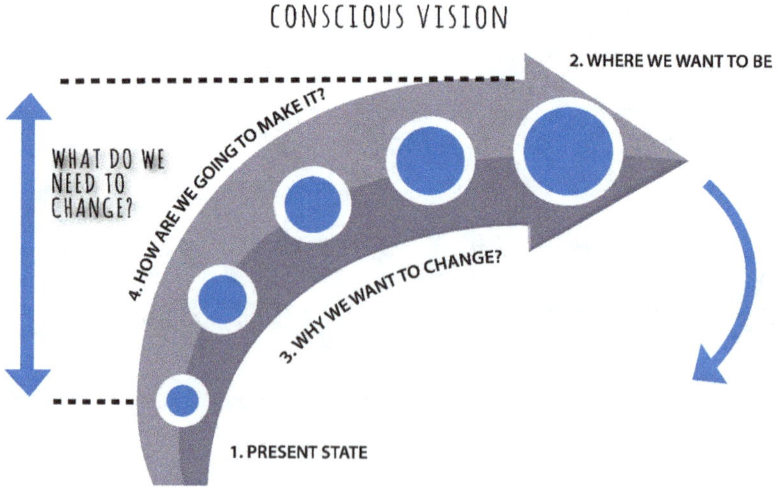

Figure 3.
Source: Portillo, E. (2016), Forces Model Adaptation.

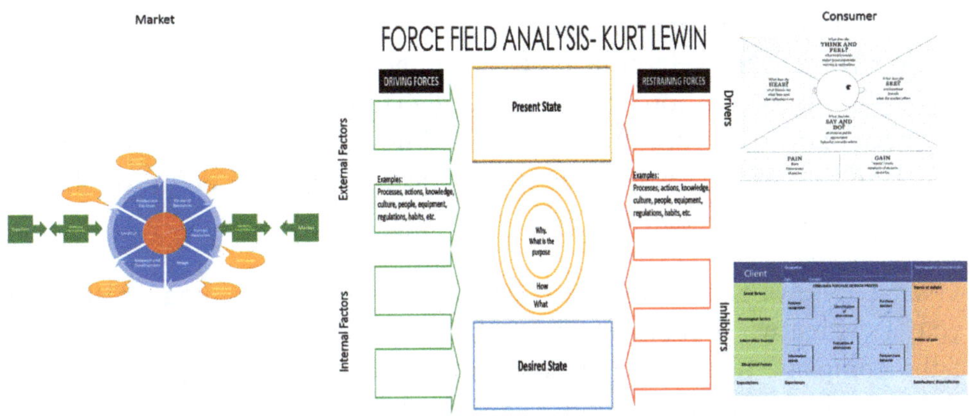

Also, in the chart above, we can see how we separate market information into two key elements: market forces (from the point of view specific to our niche market) and factors (from a generic point). This is why it is necessary to differentiate external and internal factors of the organization, as well as the drivers or inhibitors that affect the buying behavior of our customers. The graph shows three of the possible tools to use: the *market forces diagram*, the *empathy map* (Osterwalder, 2009), and our model of *Customer Journey Canvas*, leaving open the possibility to use the tool that the reader considers more appropriate for the collection and simplification of information.

STEP 9. HIGH IMPACT VARIABLES IN THE DEFINITION OF INNOVATION STRATEGIES AND PROCESSES

Having explained this, it is time to proceed to the next step: define and prioritize the processes or factors that must be strategically adjusted in our company or project.

As we have seen, the SWOT analysis (mindtoolscom 2017) allows us to gather information in a simplified way. However, it is necessary to clarify a relevant point when it comes to align the information that is poured into it with the generation of strategies within the company.

There are usually three situations we must take into account during this transition process:

1. Data generates information, so this information must generate strategies
2. Market forces, by nature, are not processes or strategies of the organization, they have to be transformed.
3. We will often face confusion because the forces of the external macro environment are forces that the organization does not control. It is necessary to ponder them and regarding their importance, transform them into processes or strategies to be developed in order to address them in the most appropriate way.

What follows is to prioritize such information, depending on the life cycle in which we find ourselves in, and the need to manage the resources of the company. If resources were infinite, there would be in no need of administration, given that resources would flow continuously.

Key Issues

Before proceeding with the alignment of forces and their transformation into processes, it is important to do another simplification exercise:

The SWOT diagram should provide what we call key issues for the transformation process of a company or innovation project. These key aspects are the ones defined as the most important ones in order to reach the defined ideal state; it is about putting in two or three words the strengths, weaknesses, opportunities and threats encountered and that we consider more important to reach our objectives. We call *Key Issues* to the simplified information.

It is recommended for prioritization and focus, to consider between 6 and 12 key elements, a smaller selection could limit the strategic vision and bigger one could hinder the implementation of the strategy due to the resources limit (time, money and effort).

As we can see in the chart below, the forces identified as key elements must also undergo another adaptation in addition to the simplification of the factors in two or three words. The forces must adapt to activities, *preferably* processes that define in a clear way what the company considers should be developed as a priority in an organization. The identification of processes is a broader strategic vision that allows the creativity of the extended teams to define strategies and precise actions to achieve the objectives within the defined processes; this is what we call innovation initiatives in our Stragile Model.

Kurt Lewin's *Force Field* diagram (Mindtoolscom, 2017) is a tool that can help us with this task. It suggests two types of forces: impelling and restrictive ones.

The forces can be defined as processes, actions, knowledge, culture, people, equipment, regulations, habits, etc. See Figure 4.

Once again, our best recommendation when doing this exercise is to think and define the forces as processes, so that, at the end of the exercise, we can link it with the identification of traditional *enabling processes* and *value-guiding processes* (the definition of these processes will be clarified later).

An Important Strategic Clarification From a Marketing Point of View

Innovation projects are generally linked to the development of new products, whether for consolidated companies, acceleration, or incubation. However, the request for support is linked to companies that want to develop "*marketing*" strategies is much more frequent.

Figure 4.
Source: Portillo, E. and Chacón, J. (2016), Forces Model Adaptation.

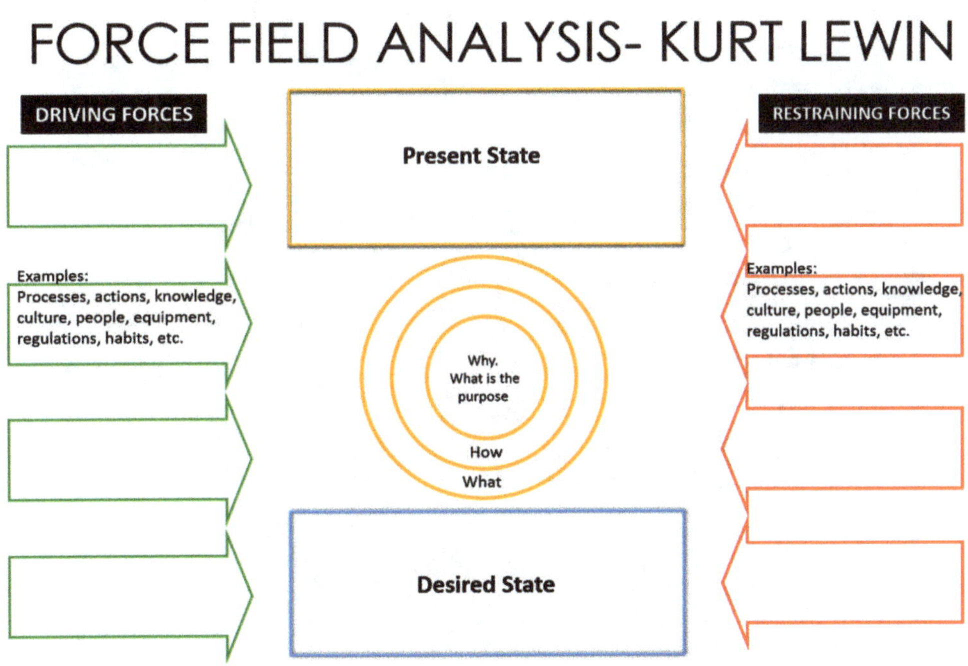

When we ask them, *what do you mean by marketing strategies?* Usually the answer is promotion! When we ask, *what do they specifically refer to with sales and promotion?* The answer is advertising!

Then comes the disillusionment:

What happens is that we are not an advertising agency! And advertising is just one element of the promotional strategy that comes to complement the brand-positioning strategy of a company. For this, before accepting any project, we must be sure that we have a product (including the price variable) that represents a true proposal of highly differentiated value for a blue ocean!

It is the moment in which both sides realize, with sadness, that the concept of marketing still has a lot of potential to be applied correctly.

Having said that, it is necessary to share one of the most relevant matrixes in marketing from a strategic point of view. The Product-Market Growth Matrix by Randall L. Schultz (Schultz, 2004).

Figure 5.
Source: Portillo, E., Chacón, J. & Moreno, T. (2016), Adaptation from: Schultz, 2004. The product market growth matrix.

It is very important to consider this matrix when defining and prioritizing the strategies to follow. This will help the users of the Stragile methodology in separating the variables that make up the value proposition (product and price); the variables that imply activities for the delivery of the value proposition (promotion, place, and service).

Often, this also helps us to clearly define the current value proposition and its attractiveness in the market. You must already imagine the reactions we have encountered when the same customer realizes that marketing goes beyond advertising, or worse yet, when it realizes that marketing strategy really begins with what we have already seen: according to Seth Godin, "*Invent stuff worth talking about*" (Godin, 2005).

STEP 10. PROCESSES PRIORITIZATION AND ORDER: THE FAVOR DIAGRAM

Identification of Forces

A bit over 20 years ago, Hammer and Champy delighted us with a masterpiece in Business Process Management. Their book *Reengineering the Corporation: a manifesto for business revolution* is, perhaps one of the first books to suggest a model of strategic adaptation under an agile method (Hammer & Champy, 1994)—a term that has been better adopted by companies In the field of information technology. But what we consider should be understood by companies in general, is to develop processes of corporate innovation that respond opportunely to the needs and adjustment of the market. As we will see later in this chapter, traditional innovation techniques and tools combined with current tools provide a useful mix of agile implementation of innovation processes for the development of new products and services.

Renewing your competitive capacity is not about making people work harder, but about learning to work differently.(Hammer & Champy, 1994)

This refers to the approach or culture of corporate innovation that must be implemented in organizations. Since then, they have suggested the importance of serving customers, competition, and above all, change. They showed the crisis as a non-existent concept; as an element integrated with the corporate routine and suggested that, far from being concerned about it, we should concern ourselves with what they called process reengineering.

In part, the basis of the proposal of this book is precisely this concept, with the inclusion of some marketing, planning, design, and entrepreneurship tools that had not been addressed 20 years ago.

We state that the base of this book is the concept of reengineering because due to the combining of traditional and current models, we have found the opportunity to develop processes and products of value in an agile and efficient way (in terms of resources) in such a way that they can be validated in short implementation periods of time before scaling them. We have seen that the "*Nail it; then Scale it*" approach (Furr & Ahlstrom, 2011) also applies to process of reengineering (and offers no discussion for the development of new products and services).

When the first stage of market analysis has been completed and we have been able to filter the information to focus on the key elements, the next step is to give them priority and focus.

The traditional models of strategic planning usually rely on brainstorming sessions (as we have already explained may be obsolete or inefficient), which manage to generate some kind of group consensus on the most voted ideas that usually end up being aligned with our current level of consciousness, so we end up suggesting; voting to develop more of the same since our fear of the changed usually exposes excuses of why we should not even turn to consider disruptive ideas.

This case is different. People are not asked to develop new ideas (since most of us do not have this disruptive vision), but to clearly identify those we consider as the most relevant forces to consider in order to generate an action plan.

Before proceeding to explain the next step, it is important to "*begin with the end in mind*":

Remember that leadership requires direction. In order to align an organization or innovation project, it is necessary to define the innovation initiatives required to achieve the desired state. Then, it follows the process of aligning the tasks within well-defined processes that eliminate the task conflict with an organizational architecture that shows clarity of roles and accurate processes.

STEP 11. THE PROCESSES MAP AND ITS RELATION TO "THE FAVOR DIAGRAM"

A map of processes, as defined by Hammer is a *Statement of Stratergy*, is a graphical visualization of an aligned organizational strategy based on the most relevant needs for an organization's clients (internal and / or external) (Hammer & Champy, 1994). Within these, three types of different processes stand out: Rector, Enabler, and Value Processes, which are defined in Figure 6.

Having this model in mind will help you understand the purpose of the following tool: The favor diagram (Figure 7).

Once the *key elements* have been defined, the next step is to align them so that we can coordinate the different strategies to be developed within a *process map*.

In the favor scheme, each key element is emptied and a process of defining relationship and impact between them begins.

The key question in order to find the relation between the variables is: *Does the variable "X" favor the development of variable "Y"?* or d*oes the variable Y favor the development of the variable X?* This way, you can try to establish the impact and relation that each variable has over all the others.

Usually, the relationship is unidirectional, in some cases it could be bidirectional, and in very few cases, there may be no relationship (although experience has told us that this rarely happens).

Figure 6.
Source: Portillo, E. & Martinez, A. (2016) Process Map.

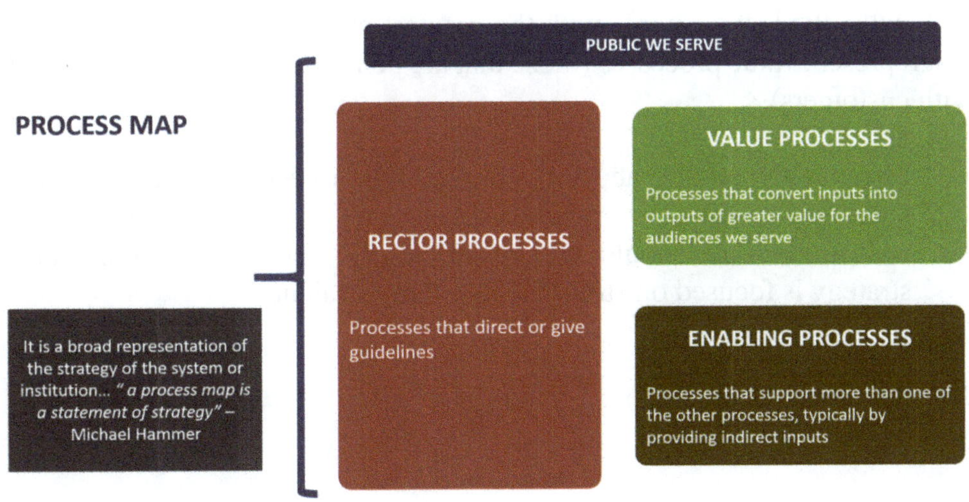

Figure 7.
Source: Portillo, E. and Martinez, A. (2016) Favor scheme.

Example

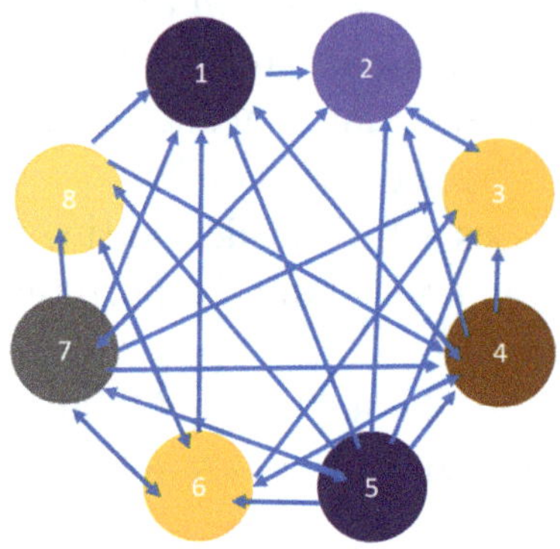

#	Name	Inputs	Outputs
1			
2			
3			
4			
5			
6			
7			
8			

After establishing the relation between each of the variables, the entries and exits that each of the variables show are marked in the box to the right.

At this point we proceed to define the possible alignment of each of the key elements in relation to the process table.

As a rule, the variables with the greatest number of outputs tend to be the guiding processes, since they practically direct/define the required course of the organization.

Variables with the same number of inputs and outputs, or whose relationships do not mark a clear trend towards a greater number of inputs or outputs, tend to be enabler processes and they receive guideline from the rector processes and develop value processes.

Finally, the key elements with the greatest number of outputs are those that represent value processes (those that are being delivered or have contact with customers).

- **Note:** When talking about customers, it is necessary to consider that some of the strategic processes defined may be focused on internal customers. This should not be surprising when the organization's strategy is focused on strengthening human talent.

STEP 12. OPERATIONALIZATION OF INNOVATION INITIATIVES

In this step, what follows is to begin the definition of goals, key metrics, dates, and responsibles of reaching the objectives defined for each initiative. The model in this sense is the same as the known models of project management.

A visual example would be as shown in Figure 8.

101 Design Methods

The reader may not know, differ, or doubt the tools suggested by this book's authors, in which case we offer two options to continue with this Stragile strategic adaptation model:

1. Contact us to discuss any doubts you may have regarding the application and approach of this proposal.
2. Select the techniques and tools that best suit you, and that you may have greater knowledge and control over.

Figure 8.
Source: Portillo, E. and Moreno, T. (2016) Project Management by initiative diagram.

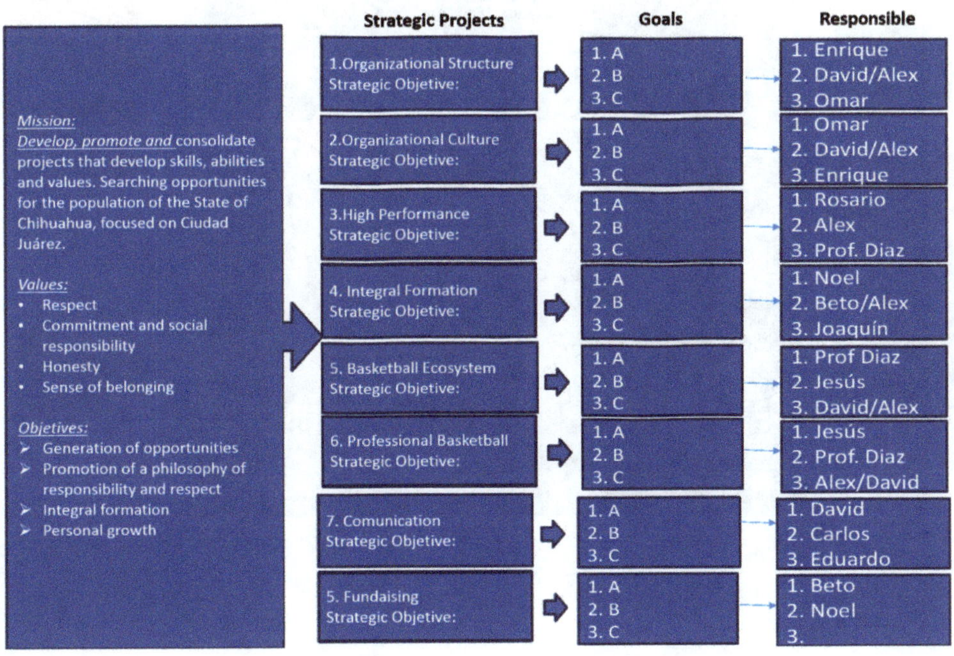

In the second case, we would like to recommend a valuable tool in the application of this Stragile strategic alignment approach: VIjay Kumar's book *101 Design Methods*.

In his work, Kumar makes a great recompilation of different techniques and tools for each one of the processes that he recognizes within the administration of business projects (Kumar, 2012). For this, Kumar defines seven stages of a business project with their respective support methodologies (Figure 9).

As a summary of his work, we present Table 1.

As you can see, the reader now has 101 tools (or methods) to select at its discretion for the stage of problem-solution and product-market adjustment.

Kumar does not get into the task of defining methodologies or tools for the escalation stage. The reason is simple, and perhaps is the same as to why we will not do it in this book either: The stage of escalation has much more to do with every single one of the tools, methodologies, and traditional management and engineering theories that have been proposed for years.

Figure 9.
Source: Moreno, T. (2017), Adaptation from:Kumar, V. (2012), 101 Design Methods.

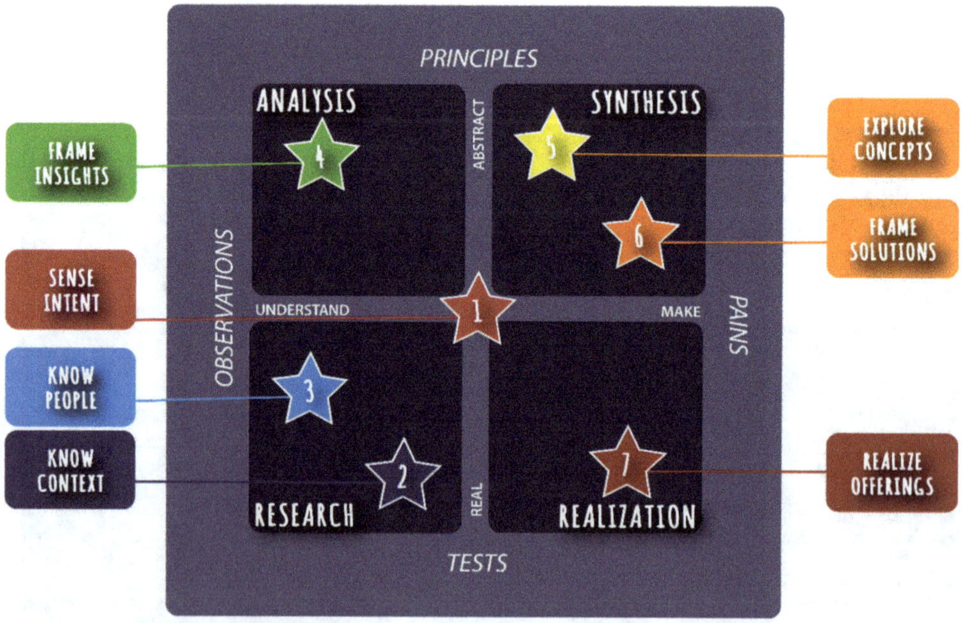

Table 1.

Stage	Goal	# Methods
1. Sense Intent	Sensitization and Megatrends	14
2. Know Context	Situational Analysis	13
3. Know People	Get to Know your Client	15
4. Frame Insights	Information Summary	20
5. Explore Concepts	Generate Alternative Solutions to Problems	17
6. Frame Solutions	Validation of Ideas	13
7. Realize offerings	Development of Products	9

STEP 13. IN-DEPTH CORPORATE INNOVATION METHODOLOGIES

We all have immersible power and capacity to reinvent our life. (Stephen Covey, 2004)

Can only few people be "inspired" to generate innovations? Alternatively, *is it possible to have a systematic method for generating innovations applicable to most companies?* Many companies want to innovate, but they do not use efficient and agile methods to innovate. They are limited to trying the method that is fashionable, but more importantly, they do not know how to deal with the complex human processes involved in decision making in the process f innovation. They waste many resources thinking they are efficient and innovative and make a lot of unnecessary spending mistakes without this leading to learning. At the end of the next two chapters, we will find an explanation and a method of innovation, tested and based on research, beyond trendy methodologies.

All human beings have a natural inclination towards creativity and innovation capabilities. Not to imply someone gives those features to us but, humans, no matter from what part of the world, have that natural tendency. This tendency is reflected in young children asking questions of various things until they find a satisfactory answer. To inquire, to explore, to experiment, how often have all heard apparently simple questions from small children, but extraordinarily difficult to answer? for example, my son recently asked me "Why are there poor people?" All human beings have that natural tendency to reason, find, and imagine solutions for everything. *What happens next?*

Why, as we grow and work, creativity and the ability to innovate become so scarce resources when solving problems or even creating new products?

It is well known that the importance and the relationship that exists between the capacity of innovation of a company and its competitiveness is a clear and logical argument. And if this is so clear, then *why innovations are not widespread across most companies and people? Why not all companies look to innovate? Do companies not want to be competitive?* It seems that it is something too complex to try and that only a few companies and people are worthy of innovation.

There are different aspects that inhibit both the creative process and innovation, and one of them has great influence on our professional behavior: the schooling. The school education that we have received the majority of the managers and employees, has a direct influence in the capacities of innovation of the person who has received this education, if as my dear reader listened to it, the education, or rather the *educated* education has been an important inhibitor for the development of creativity and innovation in people and, as a result, in companies.

The approach of this book is that people have to re-educate themselves to develop their innovation capabilities, situation which we will briefly discuss in this chapter.

We have many possibilities to look at in a blank sheet of paper, the one I prefer, is to see it as the greatest opportunity, because we can let our imagination fly in any direction and we can create new things. I have spent a good part of my life convincing people that the blank sheet of paper is the greatest opportunity in the world and is not frightening at all. Mart Sklar, Excutive Vice President (Imagineering Ambassador) Walt Disney Imagineering

Companies can only innovate through people. Innovation is a social process of interaction, but for this to happen, we must create a conductive business environment and a team of people with the knowledge and tools necessary to venture and "*risk*"that have an attitude of innovation or an innovation mindset. So, it is important to create an organizational culture where people want to be innovative. It will certainly take time, but it is possible to produce quick hits during this process.

Innovation is about how we perceive ourselves for the creative process. I believe everyone is capable of creating something new and valuable, as it is part of the human nature, it all depends on having the right mindset: the innovation mindset, as we discussed in the previous chapters. To generate an innovation is a path; it is a process of discovery. Innovators are made, not born; but they are in a constant need to develop at least four things: creativity, sense of self-efficacy, flexibility and persistence.

Creativity should be continuous in our way or thinking, but because of "*ordinary-scholarized*" education and professional life, creativity has become rare in the human brain activity. In addition, creativity in the evolution process has been useful only when environmental changes required an innovative response, and creativity requires high-energy to be accomplished.

Creative behavior can be taught, learned, and even provoked. You and your organization can train for that. Self-efficacy is defined as "self-belief in one's ability to succeed in specific situations," and can be developed with the correct human interaction during the innovation process. Self-efficacy is an issue of perception, self-perception, and the relationship between self-efficacy and entrepreneurship behavior has been statistically proven. Albert Bandura, a scientist focused on the perception of fears, found that small, discrete steps, accompanied by small-celebrated successes can lower and (in some cases) eliminate the fear response, increasing the sense of self-efficacy.

Other element involved in what we call the *innovation mindset*, is the mental flexibility. In the journey of discovery, while surpassing the obstacles during the creation of something new, you will need to constantly adapt your initial ideas and even add new ideas along the way. It is also common that during the process, you find your initial guesses to be wrong, which is part of the process and is perfectly fine. Please do not fall in love with your first ideas or hunches, but with the problem or need you want to solve, because you will need to change your ideas many times.

Persistance and learning from failure are intimately related with the *innovation mindset*, cause in the process the innovator faces a lot of obstacles and challenges, and there is now innovation innovation in the world that first doesn't face many obstacules and failures, and persistence is an essential capability to overcome those obstacules and failures, learning from it during the innovation proccess.

Then, *what is innovation?* For the purpose of this book, we want to manage a very simple but, at the same time, powerful innovation concept. We will assess innovation as a new service, product, or system that hits the market at a steady pace in a determined industry. This means innovation will depend on the market success, if not it could be an invention of something very creative, but not an innovation.

Of course, there can be cases in which a company produces an impressive and complicated technological device. If this device in not bought by clients, it is considered as creative, but not innovative. Since innovation depends on a market opportunity for whatever we've created.

The innovation process requires an entrepreneurship effort, given that without it, innovation simply would not happen. *Innovations serve or fulfill a market need.* The invention process is one thing, but a market fit may be more complicated.

Why are innovations and entrepreneurship so important? As stated in the World Economic Forum meetings (2008 to date), these are the key issues to ignite the development, generation, and sustainability of the economic growth.

Beyond the performance of a particular company, the importance of innovation as a company behavior has been stated all over the world, since this behavior translates in regional and nationwide developments reached through sustainable economic growth.

Innovation is a social ability reachable for all countries, regions, companies, and individuals. Nevertheless, if a business wants to growth constantly be profitable, the company needs to generate innovations, not only as new products or services, but by acknowledging their competition' strategic competence.

Creativity and innovation are part of the human nature, which should and could happen on a regular basis in any company. Bottom line, innovation is about a company's organizational culture, values, and beliefs.

Figure 10.
Source: Moreno, T. (2017) Based on Nathan Furr, Paul Ahlstrom: Nail it then scale it (2011).

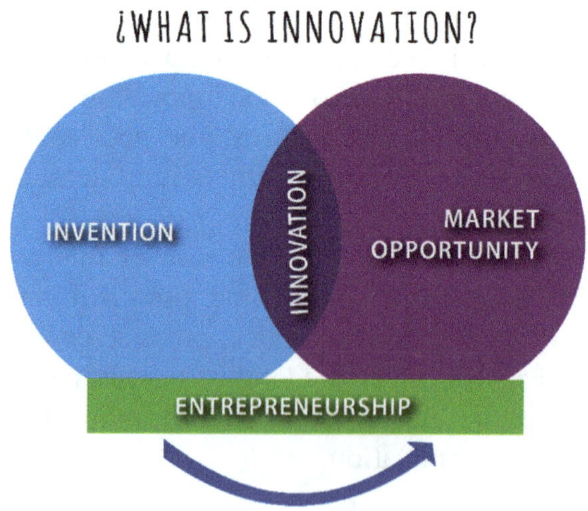

There are many reasons why companies are not innovating, but we would like to emphasize in the following:

1. Lack of entrepreneurs and entrepreneurship behavior
2. Organizational culture and structure
3. Know how (wrong innovation methods)

All three issues are interrelated.

First, innovation needs an important entrepreneurship effort; intrapreneurship initiatives take time and failed attempts. All innovations require an entrepreneur or intrapreneur. In the real world, organizational structures, culture and regular companies and habits inhibit entrepreneurship behavior. Someone who want to change something, normally is not welcome, is a natural human reaction; resistance to change the common activities. A company that really wants to produce innovations in regular bases and not just for serendipity, needs to provoke, protect, and reward intrapreneurship behavior. This behavior is result of a mindset; a way of thinking and perceiving ourselves. Innovation and entrepreneurship mindset are pals that must work together. It is possible that in companies (mainly multinational), innovations are being produced only after overcoming a lot of internal obstacles, and these innovations get to the market despite this organizational culture and structure. These means there are intrapreneurs capable of "jumping" all kinds of obstacles and company's bureaucracy. Organizations should design safe environments where employees encourage risky things, not restrictive walls.

One strong inhibitor of intrapreneurship and innovation behavior is the fear to failure. Executives, even successful ones, fear and manifest this fear through the most common form as autocratic leadership. This is or was a well-learned emotion through different "professional" experiences and social practices where failure was either punished and/or ridiculed. These experiences become a very important obstacle for innovation, because innovation requires failures and learning from failure. This concept is what we call "*The Law of Failure,"* and it is not compatible with bosses who are *always right*. Innovation is a team sport. It requires collaboration among ideas, data, information, and arguments; it is not a hierarchical procedure.

As a matter of fact, fear of failure is a justified fear, it is quite improbable for any executive to produce *the* big idea. Success rate in innovation projects is and has been traditionally low. Sometimes, innovation occurs by chance and not by a delivery process. The percentage of success of new products or services, even in multinational companies is ~20%; in start-ups, the success rate is even lower, just ~5%. Failure is costly not only for the company, but for the reputation of the person who developed the failed idea.

With these success probabilities, natural and logical social behavior dicates not to attempt innovation projects, since the chances of success are very low and the risks are much greater than the possible rewards. It seems that seeking to generate innovation is too risky and complex to try, and that only a few companies or people, perhaps people with *superpowers* are worth innovating.

Innovative process implies creativity, experimentation, failure, errors, and risk. If we want to form entrepreneurs or even innovative employees, we need an educational system (or an organizational culture) where critical and independent thinking is the norm of performance, but independent thinking is not compatible, even contradictive to the traditional educational system that unfortunately prevails in our sociey.

A Although the theme of fostering innovative behavior is subject of another book, it is needed an educational system that: nurtures creativity, capacity for change, curiosity, adventure, discovery, risk-taking; all without fear of criticism or punishment.

If an education that inhibits innovation is received, there is no culture and organizational structure oriented to encourage entrepreneurial behavior; the chances of success of innovative projects are very low and the failure is punished, we have a non-conducive combination for the generation of innovation in organizations. Unfortunately, this is the situation of most companies so it is quite understandable that innovations are not a habitual behavior, as in a business and a socially environment we settle for surviving, and keeping things as they are, because to us it has become more important to keep a job than to risk the professional career with a project that has high chances of failure.

In addition to these conditions, we companies lack the "*know-how*" or correct methods to produce innovations. Employees in most of the companies don't have adequate tools to generate innovation. They just don't know how to do it.

For instance, the normal process of innovation is represented by the "*Waterfall Development*" steps (Winston, 1970), which seems to be logical. First the idea, then the requirements, design, implementation, etc. The problem with this process is that it parts from the assumption that we have the correct idea for the market, so the whole issue is to generate the best idea. But this assumption of having the "great idea" is very dangerous because successful innovation is not about the quality of the idea, but about the quality of the process you follow.

Figure 11.
Source: Moreno, T. (2017) Based on Winston (1970) and Blank (2005).

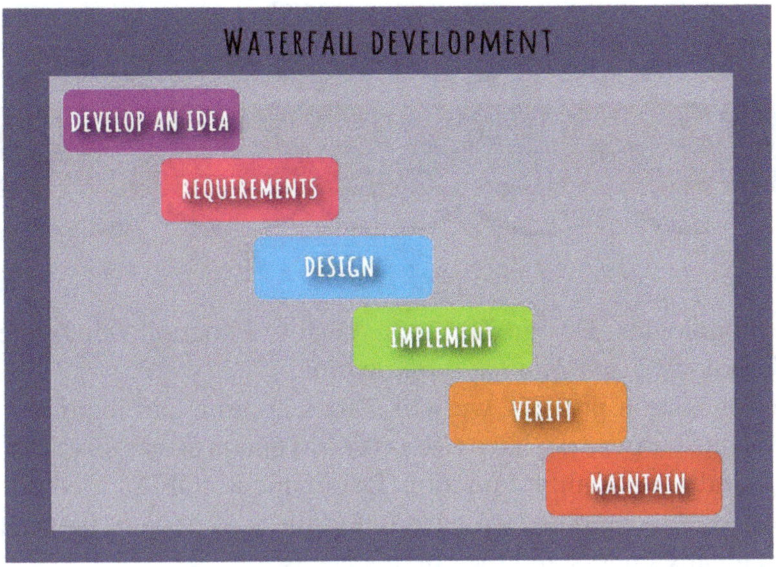

The actual innovation process is not working because we are set on a dreamland. We must land into a market reality and take into consideration the market information that will allow us to adjust these ideas and make them work.

But what if there were a set of methods, skills, and knowledge where the rate of failure is very low? Let's think about a situation opposite of what happens now, where the success rate is 80% or higher, could it be? *Is it possible to train the employees of the company so that their innovation projects have a success rate >80%?* We believe it is. We believe people can be re-trained, and that innovation can happen. We also believe this may be a reality to the vast majority of companies, micro, small, medium or large with the resources they currently have, but with the right methods and an organizational mentality and culture focused on innovation.

We propose another process centered in the market needs, following the general principles of Lean Start Up (Ries 2011) and Lean Innovation, in Figure 12.

There are a lot of methods, but if we combine some of them we can get a practical and systematic method to produce innovation. Yes, we can have innovation and structured process proved to be applicable to any industry in order to create breakthrough innovations by combining the best elements of

Figure 12.
Source: Moreno, T. (2017) Adapted from Furr, N.R & Ahlstrom, P (2011).

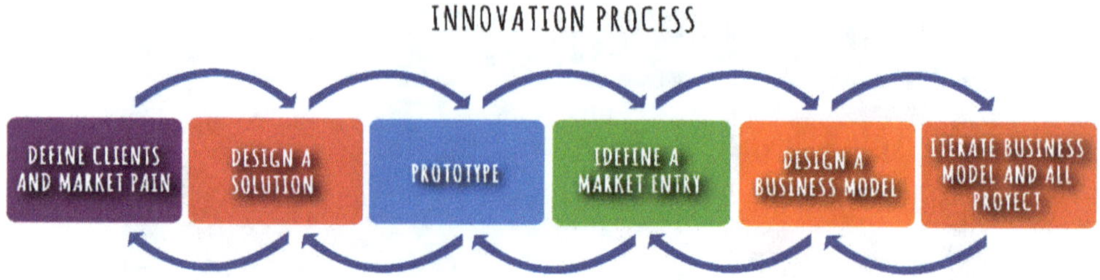

three methodologies: Design Thinking, Quality Function Deployment, and TRIZ Methology; *Lean Inteligent Innovation.*

In the innovation process, we will start searching and inquiring market needs. For this purpose we have two powerful and proven tools: The Design Thinking and the Quality Function Deployment (QFD) Methodologies. Among another techniques, we have found these to be very powerful tools to systematically understand the client's needs.

Design Thinking Breafing (DT)

Initially conceived like a "*way of thinking*" in the book of Herbert A. Simon, *The Sciences of the Artificial* (1969). Subsequently, it was presented as a design or engineering tool of Robert McKim's *Experiences of Visual Thinking* (1973). Rolf Faste expanded McKim's work in the 80's and 90's in his classes and works at Stanford University, defining and popularizing the concept of Design Thinking as a form of creative action (Fastefoundationorg, 2017).

Peter Rowe in his book *Design Thinking* (1987) first used the term within the design literature context, providing in his book a systematic explanation of problem-solving procedures used by architects and urban planners. In a 1992 article by Richard Buchanan entitled "*Evil Problems in Design Thinking*," the author expresses a broader view of design thinking centered on understanding human needs; user needs.

But what the DT project contributed was the application of this concept in the founding of IDEO by David Kelly, design professor at Stanford University. IDEO is a design and innovation firm based in Palo Alto, California in 1991. In this company, the DT methodology was applied to the design of new products. Their designs, new products including technologies from the Apple company, won numerous awards and have been exhibited at the Museum of Modern Art in New York, the Axis gallery in Tokyo, and the Museum of Design in London. Becoming probably the most important new product design company in the world.

Design Thinking is a creative methodology for the practical resolution of problems or needs, which seeks to develop solutions that are superior to traditional solutions. This concept is based on understanding the needs of the client or user in depth. It is a methodology that incorporates a philosophy of design focused on the needs of the individual. In Design Thinking, innovation is nourished by a deep understanding through direct observation of what people want and need in their lives and what they like or dislike about the way products are made, packaged, marketed, and sold. Design Thinking techniques are not only applied to physical products, but in all human-centered activities in which Design Thinking such as process design and services, or even ways in which people collaborate.

Having as a starting point a comprehensive understanding of people's needs, through empathizing with the needs of the people, observing the context of the use of the product or service in detail, seeking to identify latent needs that at first sight are not observed, and using this knowledge to inspire innovation, Design Thinking facilitates to imagine solutions that are intrinsically desirable.

A relevant element of DT is to carry out early experimentation of possible solutions instead of speculating on its feasibility, creating a rapid prototype testing culture and early experimentation. Once a highly desirable solution is encountering through interactions with people, it allows moving to the feasibility terrain. The DT proposes that the complexity of products, services, and business models has replaced the myth of the creative solitary genius by the vision of the interdisciplinary and enthusiastic collaborator. This is how DT seeks to form a team experienced in more than one discipline, fostering a lateral thinking and developing the ability to work collaboratively. For example, in IDEO, each innovation project employs people from different expertise such as doctors, engineers, marketers, anthropologists, industrial designers, architects, and psychologists.

The DT states that no matter how challenging or difficult a situation or restrictions are, there is always at least one possible solution that is better than the existing alternatives. Instead of producing *"the great idea,"* or planning based on the typical milestones of project management, proposes a process of creative discovery centered on the human and followed by iterative cycles of prototypes, testing, and improving the emerging ideas.

Figure 13.
Source: Moreno, T. (2017) Based on Stanford D. School and Designthinking.

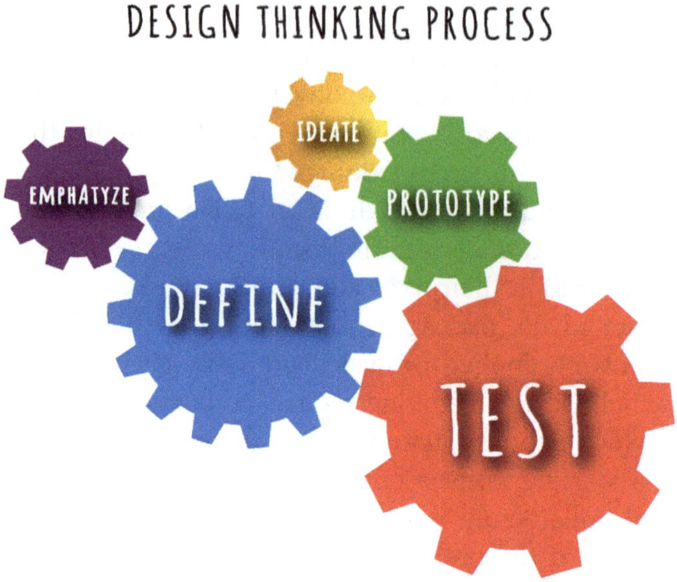

Within these five basic stages, it is possible to define and empathize with the most relevant needs of the client, by defining the problem, conceiving a solution, prototyping, and testing the ideas throughout the process. At the same time, we are looking to ask the right questions to detonate more and better ideas, and even creating opportunities to co-create with customers and consumers.

These stages are not linear; they can occur simultaneously and can be repeated. During the process, taking into account the feasibility of technology and business, innovation focuses on preferences, needs, and human behavior. Thanks to direct observation, it is easier to perceive unexpected approaches of what consumers want.

Design Thinking has an interesting concept that has been called *"evil problems."* It refers to problems that are poorly defined or unnecessarily complicated. This concept is important because a poor definition of a problem or a need to solve for a client, can lead to much wear and tear of resources generating products or services that the market does not want to buy, and life is too short to create products or services that nobody wants. In this sense, a basic principle is the correct definition of the problem, even if it takes time, because *"a well-defined problem is a half-solved problem."*

David Kelly said once, that one time someone hired IDEO for a project, and he spend at least 3 months to understand the problem, so I asked him if the clients could take that long just to understand the problem, he replied that if a customer did not agree to take the time to understand the problem, then he would not be their customer (IDEO, 2016).

So in contrast to the *"evil problems"* we now have *"domesticated problems"* or *"well-defined problems"* so from the exact understanding of the problem, the design of the solution will be feasible through the existing team expertise.

The QFD and TRIZ tools are especially helpful for the treatment of *"evil problems."* They help to structure and define them to achieve creative and impactful solutions.

If we part form an idea, most of the time our guesses will be wrong. Even with a market research, there are a lot of myths for understand market needs, one of them is market research. Among the Fortune 500 companies, only around 12% did a formal market research to introduce their initial products. (Source Inc Survey, 2008).

To deeply understand client needs, it's more reliable to perform direct-targeted observations and interviews. Through this practice, we can understand customers and users with no cost, and in a much more accurate and efficient way. The goal is to understand client needs the best we can and be able to create hypothesis from facts and assumptions, not just from *"ideas"* or *"guesses."* This is the basic need-finding process that design thinking school calls ethnographic research.

Ethnographic research is the rigorous study of the daily routine of people form a given culture in order to understanding the population's needs and points of view.

Ethnographic research starts with basic questions. Then, it understands the why's, needs, and feelings. It is important to realize that a very good source of innovation comes from understanding the why's.

We observe things with different eyes; with empathy, what people do. We observe to find client needs, we are looking for the gaps between what people do and what they need, what they value, and what they are willing to pay for. It is fundamental to understand this the best we can, the *"bridge"* between what clients have and what they need, as well as understanding the context of its use. This observation process will provide you with *"new eyes"*, that will reveal your client needs. As the poet Marcel Proust said, *"The only real voyage of discovery consists not in seeking new landscapes, but in having new eyes."*

An interview is an extended, open-ended conversation exploring needs, emotions, and client aspirations. The interview is more than data gathering. The interviewer needs to be ready to hear something new; this is important information for the design process. In the DT process, the interview is an integral part of the need-finding process. That is why innovators need to become very good listeners. During interviews, innovators look for the stories people tell about what they do and what they need while listening with empathy and looking for deep understanding of needs and emotions. A very powerful tool in an interview is the *WHY*. By the way, translated in business terms, at least in consumer goods, emotions are an extraordinary source of business.

In DT, empathy is always present during observations and interviews, and the only way to really understand the concept is by active practice.

- **Fact:** Something we know with certainty – *What is real?*
- **Inference:** Educated and logic guess to form a bridge between what we know and don't know – *What follows?*
- **Educated Assumption:** How we perceive a context based on facts and educated inferences. When we take something for granted with logical reasoning, but without sufficient proof of its certainty. *What is being taken for granted?*
- **Client Relevant Problems (Problem-Solution Fit):** One key element for innovation is to understand relevant client problems or needs. At this stage, you have to work defining relevant problems: the problem

Figure 14.
Source: Moreno, T. (2017) Adaptation from Co.Barry (2010) Anatomy of an interview.

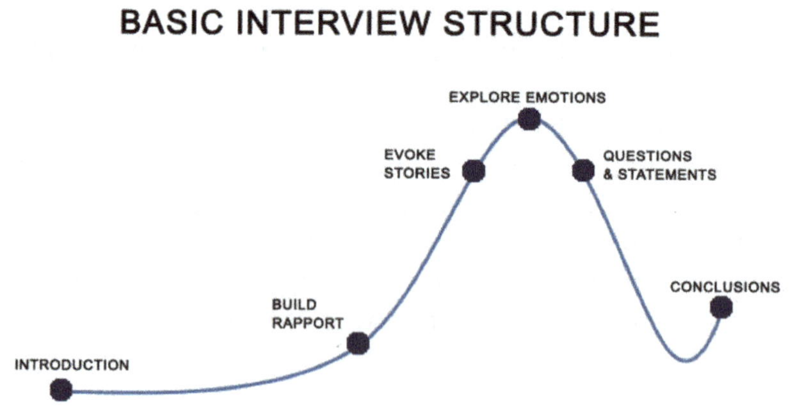

definition. A client's need is something missing within a system of use and has a meaning. Innovators are looking for these unique "missing things" that have market value. The tools are observations and interviews with empathetic language about the user, identifying a good amount of problems, translating market-field information (observations and interviews) into specific needs, articulating findings and placing them in a client context, and deciding what problems are relevant for the client or user.

In DT, once the problem has been correctly defined, the following step is to start with solution development. At this point, we must connect the concept of value proposition and business concept, tasks that require a lot of creativity and work. Designing a good VP implies to work describing you clients, define and re-define the need or problem you going to solve for them; the main features of your VP and Business Concept as a whole.

As we stated before, a real market innovation must have a market. If not, is it just an invention; a very creative idea at best, but not an innovation. It must be desirable to a group of people. It has to be feasible (possible to produce), and viable (generate cash and revenue). Desirability, feasibility, and viability is the Innovation Triangle of the Design Thinking process.

Figure 15.
Source: Portillo, E. & Moreno, T. (2012) Customer centric solution.

Figure 16.
Source: Ortiz, S. and Moreno, T. (2014) Opportunity Development.

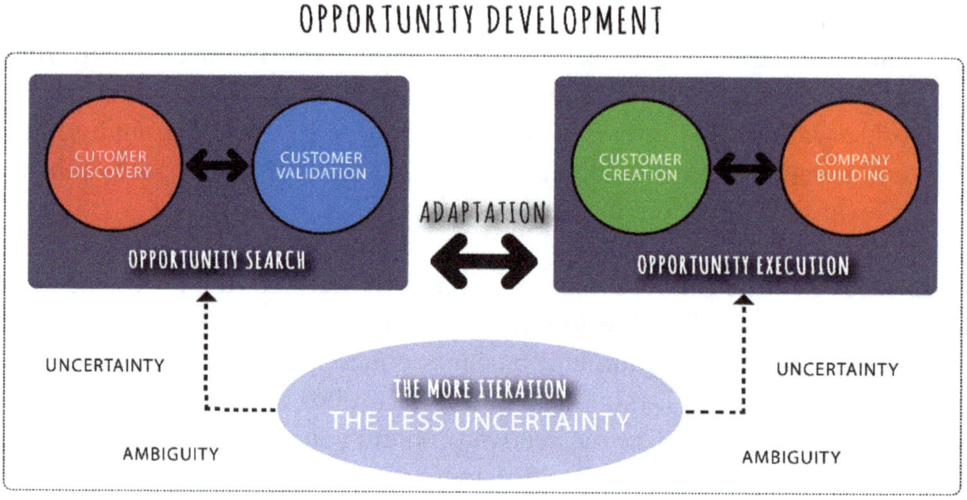

Figure 17.
Source: Ortiz, S. and Moreno, T. (2014) Adapted from Kelley & Kelley (2013).

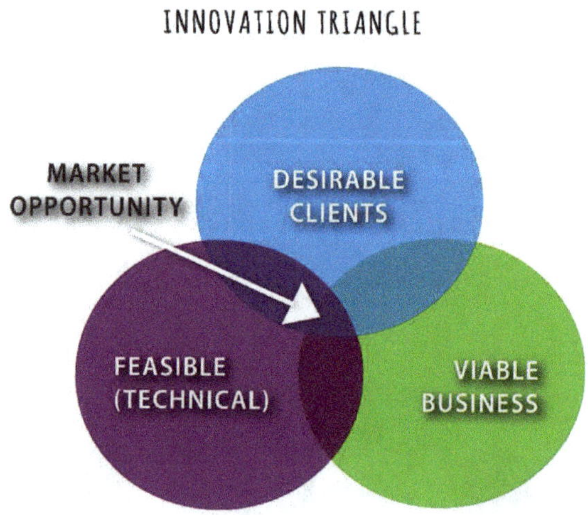

It is very important to be always client-centered. Being alert of client needs is a permanent task for word-class innovators, as well as a key-innovator activity. However, besides client needs, many innovation-successful projects are supported on the innovator interest or, even better: passions. In addition to skills and networks, if we can connect passion, skills, and client needs

Figure 18.
Source: Ortiz, S. and Moreno, T. (2014) Adapted from Byers, Dorf & Nelson (2015).

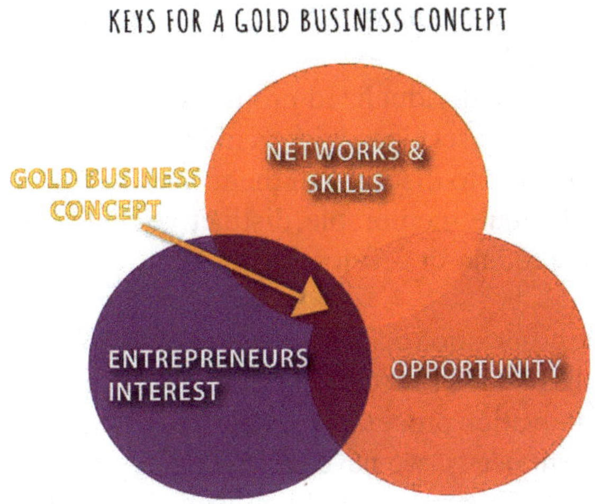

in a business concept, then we have the key elements to develop a "*gold business concept.*"

At this stage, there is a viable option to start with the value proposition development. The drivers of a market opportunity and value proposition are:

- **Desire/Need:** Personal desires, business desires, entertainment, social status, self-image, etc.
- **Pain/Claim:** Inefficiency, unreliability, inconsistency, uncertainty, uncomfortable, fear of losing out.
- **Improvements:** Saving time, saving money, quality, and convenience.

And the possible specific elements of your value proposition are:

1. **Client Convenience:** Simpler, creative and productive solutions.
2. **Price:** Cheaper similar solutions.
3. **Cost:** Products or services reducing client cost in a specific and relevant task.
4. **Design:** Products or services with better performance (disrupted increases in performance).
5. **Risk:** Clients reduce risk using the service or good provided.
6. **New:** Satisfaction of needs not previously attended.

7. **Accessibility:** Technology or service available for a specific segment market with no previous access.
8. **Design:** Functional products with superior aesthetic or more attractive than the actual solutions.
9. **Personalization:** Individualized products or services for each client.
10. **Status:** Products or services with visible characteristics of attraction, allowing for people to distinguish them.
11. **Experience:** Unique, memorable, distinctive client experiences through the use of the product or service.

This is the value proposition concept, but for a more robust value proposition development, we have the QFD (Quality Function Deployment) method that can enhance the innovation process. After the need-finding process where we deeply identify client needs, we have the possibility to use QFD, which is a power tool to prioritize the relevant client needs in a structured way, which, for the purpose of this book, we will only describe briefly.

QFD Quality Function Deployment

Change has now become an integral part of stability: today we can achieve stability only by accepting change as a continuous and permanent state. (Daniel Burrus, 2017)

Dr. Yoji Akao, creator of the QFD concept, experimented with problem-solving algorithms and the Ishikawa Fish Bone diagram to designing products, and led to a method represented by a matrix that he called: "*The House of Quality (HOQ).*" (Qfdiorg, 2017). In this matrix, it is possible to identify in a very structured way, the key elements that impact customer satisfaction. This tool matches perfectly with the ambiguity of DT process at this stage, and goes beyond brainstorming. In QFD methodology, these customer satisfaction elements, (client needs in DT) are translated into specific parameters and characteristics in new product design. The HOQ presents in a practical and structural way: client needs, specific design and engineering parameters, and specific actions.

The QFD assumes that all products and services require a process of continuous improvement to be able to compete in the market. Furthermore, similar to the DT, the QFD is in search for this continuous improvement in a structured interaction with customers.

One of the strengths of QFD is the ability to help find areas of potential improvement through prioritizing customer needs. By analyzing customer survey data and correlating that data with product parameters and characteristics, QFD helps to highlight inconsistencies between the customer's opinion of your product or service and the internal opinion of the company. The QFD was designed to generate innovations in existing companies, not only for improvement or product development, but also to improve internal processes by identifying internal customer requirements within the company, thus improving satisfaction and efficiency of the company.

QFD is a proven and structured methodology for continuous improvement or design of products and services, as it focuses on the most important aspects and identifies them more toughly than DT. The strategic objectives of QFD are to reduce costs and reduce time to market, which is probably the most valuable benefit to the company.

The QFD begins with an analysis of how customers perceive the current products of a company. From these perceptions, the list of *what* the customer perceives is derived. This list of consumer requirements is called "*Consumer Voice,*" (VOC's). Based on the voice of the consumer and with the experience of the company, the functional parameters of the product are identified.

It assigns a value to the relationship between what the customer perceives and the parameters that the product has, as a way to approach the customer. The QFD is the appropriate tool to clearly visualize the needs of the customers, as well as the evolution of the technical characteristics of the competition's products.

According to QFD, every company should go to the "*GEMBA*", where "*Gem*" means, "*present*", and "*Ba*" means "*place.*" GEMBA represents the true source of information. It is the place and time where the product is used and where the value for the customer is born.

QFD has four general phases and objectives:

1. Performance Requirements
2. Product Design
3. Process Design
4. Process Control

The main tool for the innovation process in QFD is a matrix called the "*House of Quality.*" In this matrix, the client's expectations are expressed qualitatively (the customer's voice) and are related to a set of quality parameters or characteristics (including functional requirements, restrictions, etc.).

The first step in the QFD methodology process is to identify the "*customer's wishes*" or CTQs, which is subjective/qualitative information that reflects the customer's requirements. It is the perception that customers have that leads them to prefer one product over the rest. A difference with respect to the DT is that in QFD, these perceptions are obtained by observation and analysis rather than by surveys and/or questions.

With informative purposes, we will generally describe "*House of Quality*." The complete diagram of this matrix has eight fields; each represents a different phase of product planning, which together seem to represent a house (Figure 19).

House of Quality

- The CTQs (customer's wishes) are entered in field 1. At this stage, the level of importance of each CTQ is also plotted on a scale from 1 to 5.
- Subsequently, engineering and product design experts must translate these wishes into specific parameters of the current product that are identified with the CTQs. These parameters are entered in field 2.

Figure 19.
Source: Ortiz, S. and Moreno, T. (2017).

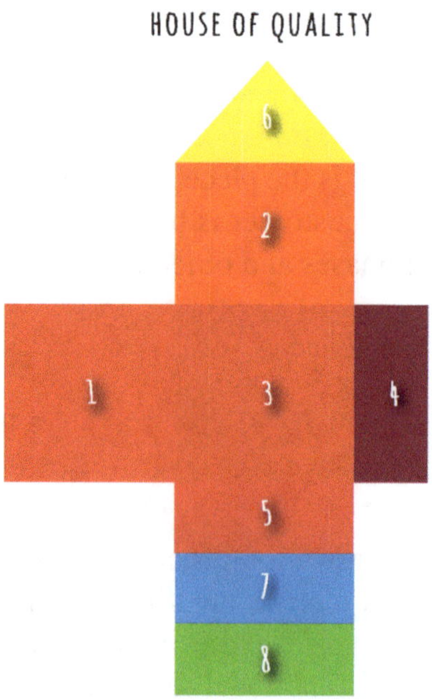

- Then all the relationships between the customer's expectations and the engineering parameters in the matrix are identified and entered in field 3.

Field 4 contains the comparison between owned product and other competing products based on customer perception.

- Field 7 contains the comparison between the functional parameters of the owned product and other competing products based on objective engineering measurements.
- Field 8 contains the relative importance of each parameter.

Fields 4 and 8 are the basis for the analysis of opportunities for improvement and innovation.

In this methodology, the user has to identify the engineering parameters of the product in question that are related to the wishes of the customer, which can be identified in field 3. Thus, observing the relative importance of the engineering parameters and setting targets of these parameters, either to match or exceed the performance of competing products (or any improvement criteria you require). In this way, what can be generated in QFD is known as "*new value curve*" identifying that it can be reduced, eliminated that which is not valued by the customer and at the same time, enhance or create the characteristics of the product or service that the customers and the market value.

Once we have identified specific clients' desires and parameters, we have an invention problem, and the best methodology to solve complex problems is the Inventive Problem-Solution Theory (TRIZ) (*Trizconsultingcom, 2017*).

Inventive Problem-Solution Theory (TRIZ)

TRIZ is a methodology that was originally developed in the former Soviet Union to support innovation processes. It was not designed to address trivial problems or incremental solutions. TRIZ is a methodology for the solution of complex problems that began with the application in the field of Engineering (in the development of new products), but then applied to any complex problem. It is based in the work of great inventors and world patents, and although it seems incredible, as a method of analysis used the principles of Marx's dialectic, especially the principles of contradiction and the negation that used to modify reality.

Figure 20.
Source: Moreno, T. (2017) Based on Dr. Noel Leon Rovira 2016 IM&ST Company.

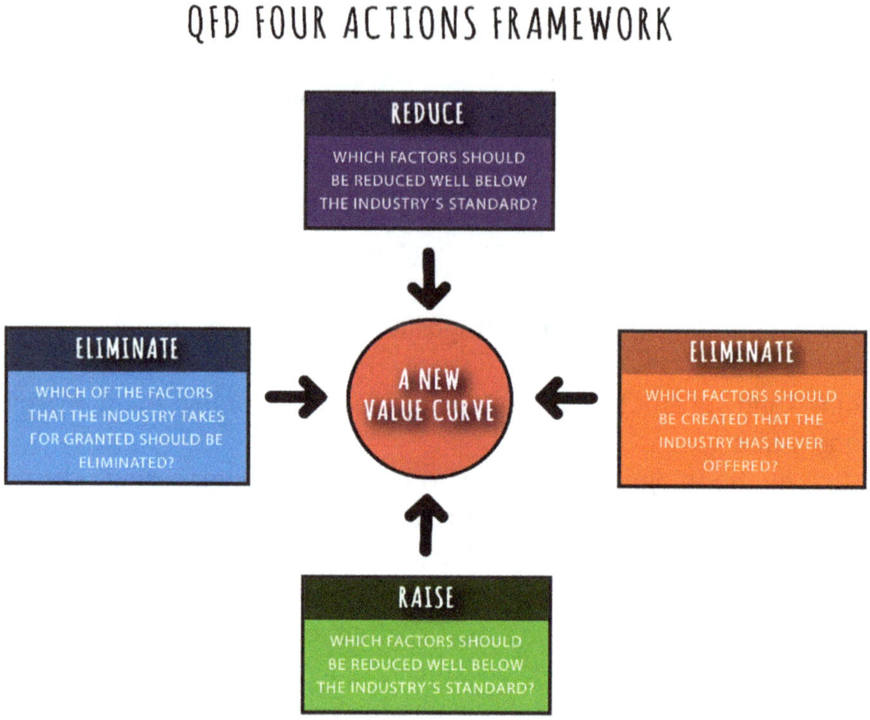

TRIZ is the Russian acronym for *Теория Решения Изобретательских Задачque* that, translated into English, means Inventive Problem-Solution Theory. The author or founder of this methodology is Genrich Altshuller, born in Baku, in the former Soviet Union, who by *"thinking differently"* was confined to Siberia from 1948 to 1954.

It was known in the West thanks to the opening of Perestroika in the 1990s and rapidly expanded throughout the world. Since then, it has been applied to other fields such as administration, medicine, and education. This methodology has even been used in the design of electoral campaigns in Russia and Israel with great success.

In 1984, the first book of Genrich Altshuller, *Creativity as an Exact Science* was published. In 1991, Altshuller's collaborators moved to other countries to spread TRIZ, presenting it to the United States. Subsequently, during 1993 this revolutionary innovation methodology (TRIZ) was expanded worldwide.

Altshuller classified innovations at certain degrees of change from elementary solutions to discoveries that transform industry and the market.

Figure 21.
Source: Moreno, T. (2017) Based on Leon, N. (2016) IM&ST Company.

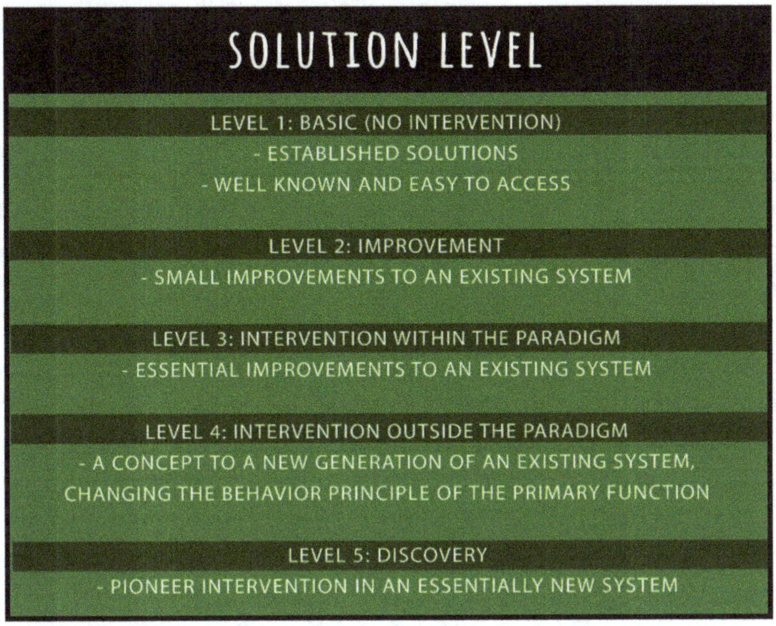

Looking to the process of how innovations occur, Altshuller found that for an "*elemental solution*" 10 errors are committed on average, but for a true discovery that revolutionizes the market, approximately 1,000,000 errors are committed. These results made him realized that the approach to making a pioneering invention of a system that revolutionizes reality, pure experimentation is an inadequate approximation.

For Altshuller, innovation should not be *almost* accidental as a process of innovation by experimentation requires. It is also impractical that the innovations are generated by chance, inspiration, or pure experimentation. So, he created a structured methodology for innovation that allows us a disciplined way to examine the problems of innovation; exploring the space of solutions generating creative ideas.

Seeking to generate an orderly process to develop innovations in a systematic and deliberate way, TRIS methodology was created. A structured methodology for innovation that allows us to examine the problems of innovation, exploring the space of application of solutions, and generating creative ideas.

One of the premises of TRIZ is that most inventive problems or complex problems have already been solved using principles from other branches of technology, science, or human knowledge. In this sense, many innovations are based on principles outside the field where they occurred.

The TRIZ methodology is based on the following premises:

1. The real problems are those that contain a contradiction.
2. Solutions can be sorted over so that they can be easily accessed.
3. Technological systems do not evolve at random, but follow certain laws of evolution.

Besides intuition, TRIZ is a problem-solving method based on logic and data, which accelerates the project team's ability to solve these problems creatively. TRIZ also provides repeatability, predictability, and reliability to the creation of solutions, due to its structure and algorithmic approach. This is an international innovation method based on science of creativity that relies on the study of the patterns of problems and solutions. To create the TRIZ methodology, more than three million patents have been analyzed; in which the main patterns that predict breakthrough innovations were discover.

TRIZ research began with the hypothesis that there are universal principles of creativity that are the basis to advanced technology. If these principles could be identified and codified, they could be taught to people to make the process of creativity more predictable. The short version of this is:

Someone somewhere has already solved this problem (or one very similar) Creativity is about finding that solution and adapting it to this particular problem.

Much of the practice of this methodology consists on learning these repeating patterns of problems-solutions, patterns of technical evolution, methods of using scientific effects, and then applying the general TRIZ patterns to the specific situation that the developer confronts.

REFERENCES

Altshuller, G (1984). *Creativity as an Exact Science*. Academic Press.

Blank, S. (2005). The Four Steps to the Epiphany (1st ed.). Academic Press.

Buchanan, R (1992). *Evil Problems in Design Thinking*. Academic Press.

Burrus. (2017). *Daniel Burrus*. Available from: https://www.burrus.com/2017/04/agility-levels-playing-field-anticipation-changes-game/

Covey, S.R. (2004). *The 7 Habits of Highly Effective People: Powerful Lessons in Personal Change*. Academic Press.

Fastefoundationorg. (2017). *Fastefoundationorg*. Available from: http://www.fastefoundation.org/publications/

Furr, N. R., & Ahlstrom, P. (2011). Nail It then Scale It: The Entrepreneur's Guide to Creating and Managing Breakthrough Innovation (1st ed.). Academic Press.

Godin, S. (2005). *All marketers are liars* (1st ed.). Portfolio.

Hammer, M., & Champy, J. (1994). *Reengineering the Corporation: A Manifesto for Business Revolution* (1st ed.). Harvard Business School Press.

KelleyD. (2016). *IDEO*. Available from: https://www.ideo.com/people/david-kelley

Kelley, D., & Kelley, T. (2013). Creative Confidence: Unleashing the Creative Potential Within Us All (1st ed.). Academic Press.

Kumar, V. (2012). *101 Design Methods: A Structured Approach for Driving Innovation in Your Organization* (1st ed.). John Wiley & Sons Inc.

Mckim, R.F (1973). *Experiences of Visual Thinking*. Academic Press.

Mindtoolscom. (2017). *Mindtoolscom*. Available from: https://www.mindtools.com/pages/article/newTMC_05.htm

Morales, C. (2017). *Forbes Mexico*. Available from: https://www.forbes.com.mx/big-data-y-neuromarketing-asi-es-como-expedia-quiere-vencer-a-airbnb/

Myers, T., Dorf, R., & Nelson, A. (2015). *Technology Ventures: From Idea to Enterprise* (3rd ed.). Mc Graw Hill Education.

Osterwalder, A. (2009). *Business Model Generation: A Handbook for Visionaries* (1st ed.). Game Changers, and Challengers.

Qfdiorg. (2017). Available from: http://www.qfdi.org/what_is_qfd/who_is_dr_akao.htm

Ries, E. (2011). Lean Startup. Crown Publishing Group.

Rowe, P. (1987). *Design Thinking*. Academic Press.

Royzen. (2017). *Trizconsultingcom.* Available from: http://www.trizconsulting.com/

Schultz, R. (2004). A note on the product-market growth matrix. University of Iowa.

Simon, H. (1969). *The Sciences of the Artificial.* Academic Press.

Stanfordedu. (2017). *Stanfordedu.* Available from: https://dschool.stanford.edu/

Winston, W. (1970). *Managing the development of large software systems.* Academic Press.

KEY TERMS AND DEFINITIONS

Key Aspects/Elements: The most important aspects in order to reach the defined ideal state.

Self-Efficacy: Self-belief in one's ability to succeed in specific situations.

Chapter 4
Integrating Human Talent

ABSTRACT

This chapter deals with the integration of high-performance work teams aligned to the structure defined by the leading team based on the inclusion of the staff. For this it is necessary to consider and implement concepts of leadership, collaborative work, and corporate decision-making processes, selection, and evaluation of personnel; but, above all, it is necessary to develop and implement appropriately concepts such as coaching and empowerment. The key to the success of the agile corporate innovation strategy lies precisely in this pillar, the empowerment of human talent. The literature regarding the topics of leadership, coaching, mindfulness, and empowerment is coarse and powerful; these are the central themes to support the launch of agile corporate innovation initiatives. This chapter explores the integration of human talent.

INTRODUCTION

The illiterates of the 21st century will not be those who know not how to read and write, but those who can not learn and unlearn and relearn. (Toffler, 1970)

Now it's time to integrate human talent! LEADERSHIP!

DOI: 10.4018/978-1-5225-4023-6.ch004

STEP 14. THE IMPORTANCE OF LEADERSHIP IN THE ORGANIZATION

The irony is that now for companies that hired employees so "*well-rounded*", innovation is a must, and without proper training, it seems that this process has neither a start nor an end. Normally, neither managers nor employees have been trained to be innovative. In fact, the percentage of companies that are trying to innovate, are few, and also with a high degree of failure. It seems that only the notables who have graduated from the Ivy Leage universities have the "inspiration" and the knowledge to generate innovations within the companies. However, the reality has shown that these prestigious few have a large percentage of errors of judgment in both administrative processes and an innovation process. Several graduates of the top universities took the decisions that led to the mortgage crisis of 2008 and 2009.

The good news is that any company and/or individual can develop a capacity for innovation beyond its current limits. If determination is present,

Figure 1.
Source: Portillo, E. (2016), The Stragile Conscious Corporate Innovation Model.

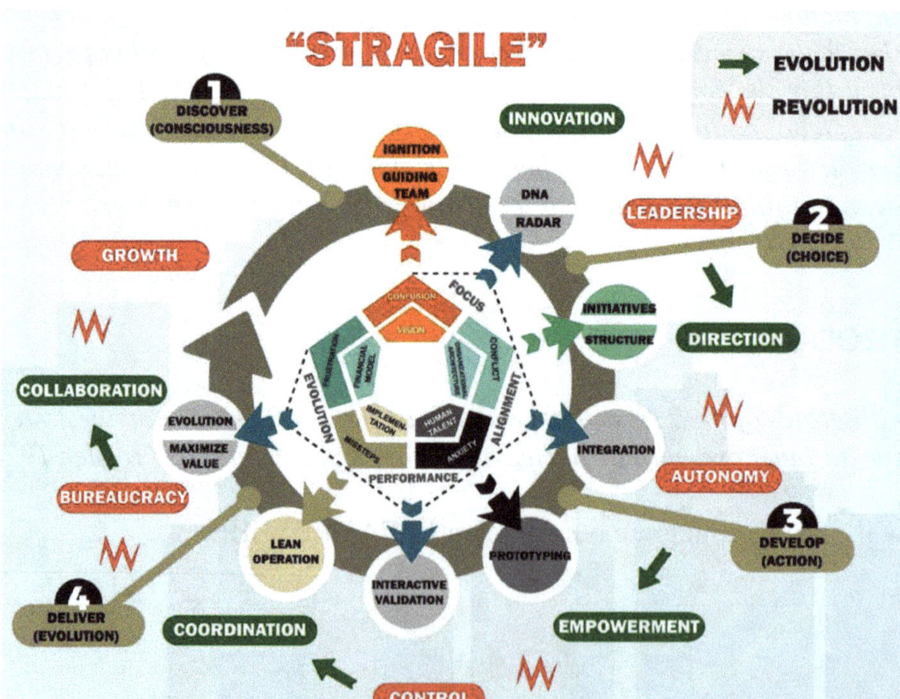

it is perfectly possible, but you have to pay a price. You have to train people, you have to re-train individuales. Like with any competency, the innovative capacity can easily be taught. The knowledge, methods, resources, and technology exists, as well as specific skills and attitudes to be developed, among which are: identification of opportunities, collaborative work, leadership, communication skills, empathy, learning from failure, facing ambiguity and uncertainty, sense of purpose, and self-efficacy; traits that were discussed in the first chapters, and given the importance of one of these, we will talk now about collaborative work and decision making in the company as key factors in the process of business innovation.

Innovation and Decision Making in Organizations

In a normal process with a logical sense of decision-making, data is recollected. With this data, information relevant to the objective or problem to be addressed is produced. Based on this information, we can establish assumptions as key tools for analysis and use different decision models that serve to guide an organizational decision properly. Currently, many mathematical models and other different types are available on the internet and literature; models that are truly accessible to virtually any manager. However, the logic remains the same: Data produces information, information is an input to generate assumptions, and either with analysis of modern methods or not, data, information, assumptions and analysis, finally an appropriate decision is reached. In fact, this model seems so logical, it almost looks infallible to make decisions most of the time. Napoleon Bonaparte was known for his ability to evaluate (make assumption) the information available and make the right decisions on the battlefield.

However, *why did brilliant people with great experience, excellent preparation and knowledge, having done incredible feats, complicated*

Figure 2.
Source: Ortiz, S. & Moreno, T. (2017).

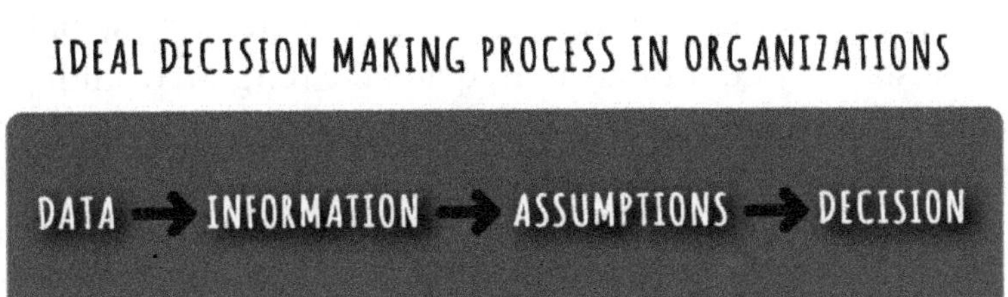

decision-making experts, have made such catastrophic decisions? Here are some examples:

The Napoleonic invasion of Russia not only provoked the fall of the Napoleonic Empire, but also the death through starvation and cold of the soldiers who survived the attacks of the Russian guerrillas. After the Chinese Communist Revolution, the decision to put peasants to produce steel in Mao Zedong's "*Great Leap-Forward*" strategy, which caused the greatest famine in the history of mankind. Between 15 and 30 million people died of hunger in a period of 3 years. Or the invasion of Iraq, not only caused a number of innocent deaths, but also instability in the Middle East and in the world that we still do not know the consequences, as well as a gap in the already large deficit of public spending in the US.

Do you believe that there were no staff members of Napoleon who did not know about the overwhelming risks of the Russian winter and even commented on them? Was there no one in Mao Zedong's body politic who realized the foolishness of putting inexperienced peasants to make steel and stop producing food? Where steel finally did not serve any industrial purpose, because of its poor quality, it is clear that in these cases, especially the one of the general and guerrilla Mao Zedong, to express a contradictory opinion meant the death and the penury of the family. In the case of the invasion of Iraq, for example, the objectives were to stop the weapons of mass destruction and, at the same time, destroy the connections or alliance of then-president Saddam Hussein with the terrorist group Al Qaeda. However, the facts tell us that there was no evidence of weapons of mass destruction and no connection between the Iraqi leader and the terrorist group. There were many voices and reports stating that Iraq posed no threat to US security, and not only was there no evidence of a relationship with the terrorist group, but the information available indicated that Saddam Hussein did not believe in a religious government. Reason why, it was considered an enemy of the extremist organization Al Qaeda. In fact, Saddam Hussein served as a center for containment of religious extremist groups in the Middle East, thereby indirectly serving American interests. The independent information and analysis available, including that of the UN was oriented to the occupation of a fragile and unstable relationship that did not attack the United States. The high probabilities of generating social, political, and military instability conditions in Iraq and the religion resulting from the power vacuum provoked by the invasion were noticed by several specialists. However, we know the outcome of the decision taken in this regard.

These examples show the making of preconceived decisions without taking into account the available information, and with ill-conceived assumptions that did not reflect reality. In this type of cases, a logic of thought is reflected in where the verdict is first performed and then the trial, leading to extreme cases of *"inventing evidence that makes the boss feel good*", in a deliberate process to justify the preconceived verdict and not the most appropriate decision, and why not say, decisions based on ethical principles.

One might think that this level of error of judgment in decision-making only occurs in the history of governments obsessed with power, but we see some examples in the business field, which although they were not so catastrophic in terms of human lives, they represented serious misjudgement that impacted the future viability of many companies, or even causing national and/or global economic crises. In the least catastrophic cases big business opportunities were lost due to having preconceived assumptions that led to serious errors of judgement. Let's look at some examples. Who would say of the Western Union President's opinion when, at the beginning of the 20th Century, company shareholders asked for a technology called "*telephone",* invented by Alexander Graham Bell (check inventor). His response was something like: "*While we think it is interesting curiosity, there is never going to be a market for that technology and, therefore, we have to decline the offer to take the licence.*" Or what do you think of the well-known decision to launch the Challenger shuttle despite warnings from several engineers about the effect that low temperatures could have on engine seals?

Or to put together investment portfolios with junk mortgage bonds, which were known to be priceless, to increase the paper earnings of mutual funds "*to make them look good",* causing one of the worst financial crises in modern history. Or the decision of American automakers, in the midst of the full subprime mortgage crisis, to produce expensive cars with expensive maintenance instead of the repeated market signals that were looking for something else. Or the rejection of the now-defunct Blockbuster company for the acquisition of Netflix, arguing that it was much risk for the Blockbuster brand to acquire an unknown company. Or more recently, companies that are making history as is the case of "*Uber*" that practically went unnoticed by all the "*investment experts*" promising companies searchers from either Silicon Valley or New York, knowing the new businesses that are going to work, overlooked, if not completely unnoticed cases like Uber, or Yahoo, in their investment funds.

Having talked about these examples that could be thought to happen in large companies and that this is just an academic discussion, put yourself in the organization where you work or some organization where you have worked. Haven't you been involved in meetings where energy is concentrated on what the boss wants to hear and approve instead of addressing the problem to be solved? Or have you not heard phrases such as *"I am the boss, is it clear?"* Probably more than once. The problem is not important, but to demonstrate the effects of the hierarchy is. While this is unfortunately very common in organizations, whether with MBA executives graduating from the world's top universities or with executives without academic degrees, it is very common for organizations to make a preconceived decision-making process. The point is that in this *"process"*, the organizational capacity to make decisions based on rational and methodological analyzes that support the best possible decisions is lost.

In spite of all the methods and what has been written of decision-making, in practice, this process is affected. Sometimes without realizing it by different emotions, within the most detrimental is the ego and the pride, since a lone senior manager comes up with great ideas. However, while organizational change towards a culture of innovation comes from above, innovations can be produced from below. Motivated people have virtually infinite capabilities to innovate.

Innovation is not generated with preconceived decisions, but with a real work group that is based on cognitive discussions based on arguments and not on emotional discussions based on hierarchy or power groups.

In all organizations, a great number of "*irrational*" hierarchical decisions are produced with inadequate or out-of-context criteria for making decisions in the company. What I call the "*Law of Irrationality.*" In basic terms, is the description of people in organizations that routinely and consistently make a lot of irrational decisions, including for personal reasons. Unreasonable decisions are made every day, for example, we all know that junk food is harmful, but routinely millions and millions of people consume this type of food.

But returning to business, every day illogical decisions are made, regardless of the circumstances. Either under the pretext of complying with policies or without this pretext, but irrational decisions are made, regardless of going against stated and objective values, trategic objectives, or even the financial performance of the company. These decision-making criteria I call the "*Business Irrationality Law.*" Despite this law, companies can advance in their goals, but they do it with extraordinary personal and emotional wear, as if they stepped on the gas and the brake at the same time on everything they do; unnecessarily wearing down the machine. Organizations tend to endure these irrational criteria usually based on hierarchy, but when irrational decisions accumulate, the company goes into crisis, sometimes crisis from which they can not recover.

Dear reader, ask yourself what would happen in the organizations if the employees were left free to design, invent the most appropriate decisions for the company without having to think first about what the boss wants to hear? It would be logical, right?

However, common sense is sometimes not used, and it is very common that a decision is made first and then, all human and managerial intelligence available is spent to look for "*justifications*" that support the preconceived and previously taken decision, to focus on the evidence the boss wants to hear. Ordinary and extraordinary meetings, and decision-making processes used a lot of energy which is focuses mainly on identifying what the boss would like to hear and then finding evidence and arguments that support the leader's decision. Under a mask of apparent rationality and accommodating facts, preconceived decisions are justified. *Why are these decision-making scenarios so common?*

The Law of Irrationality, or errors in decision-making, taking the works of the brothers Oliver Sibony, are due mainly to two aspects:

1. Of the person
2. Of the process

Let's talk first about the aspects related to the person. It turns out that decision makers in organizations are humans with emotions. We are not talking about people necessarily being over-sensitive to collaborators, but having more dangerous emotions like ego, overconfidence, or even fear of failure, which produce stress situations that are very difficult to control. Managers tend to see only information that is consistent with their personal beliefs or the decision they consider most appropriate; minimizing or discarding information that does not fit their judgment or opinion, this is related to the "*ego*" or business status where the manager is right and has to "*defend*" its projects instead of discussing them, since recognizing doubt or uncertainty is not an image that managers like to give. Some go as far ridiculing opinions different from them, because when they have arrived at a certain moment in their careers, they feel the need to portray a knowledgeable demeanor as masters of the secrets of success. Now, we understand better why Napoleon, dismissed the arguments against invading Russia after winning the battle of Borodino and achieve the withdrawal of the Russian army, with a force of 700,000 men, and the Czar with less than 200,000, but in geographical and logistic conditions that they disregared. Or the investment houses which declared that as the mortgages had sustained the economy and the country's financial development, its falling mortgage bond drop was impossible. To this method or decision-making criterion with personal biases where hierarchy is exercised as the main criterion of decision-making I call "*Management Mechanism.*"

There are several emotional aspects that can seriously affect managerial judgment; two of which are especially dangerous and harmful: Pride and ego—which are related to the need for status. These attitudes are the main source of the syndrome of managerial mechanism, which leads to the conclusion that the boss is the one who always makes the best decisions. That is the reason why he's there, that is why he gets paid; for that study in X or Y university, titles that give false status and criteria to make the right decisions without taking advantage of the talents of its collaborators. This type of manager, without the humility to learn from others and from the context, can be and sometimes is a real danger to the organization and to the process of innovation, obstacles practically insurmountable. These attitudes are unfortunately more common than one might think of graduates of the best universities in the world, since decision-making and innovation go hand in hand with the ability to observe and actively listen, and that only occurs with the actual intention of search, of experimentation, and of humility to learn. There is another aspect of the person that leads to conformism and indecision. At the time of taking few risks, where innovation does not flourish, because innovation attempts usually have a high failure rate, many companies survive because they have strong structures, but no organization can endure a permanent prolongation of the syndrome of directive mechanism.

As we have seen, sovereignty and ego are the main enemy for directive decision making. However, emotions can not be closed down. Instead, they should be present and controlled. In other words, dear reader, the main enemy for making business (or personal) decisions is yourself. Of course, you can also become your best ally for decision-making, the antidote is controlling emotions and recognizing the interests of the people involved or, at the very least, be aware of them. In innovation projects, the syndrome of managerial messianism, where the personality of the involved tends to be right at all times without being open to rational debates, is a great inhibitor of innovation processes of any kind.

Figure 3.
Source: Ortiz, S. & Moreno, T. (2017).

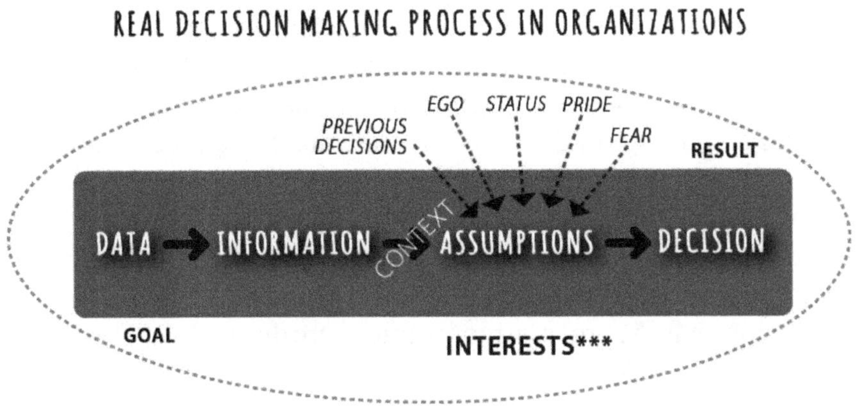

STEP 15. THE EXPANSION OF THE COLLABORATIVE FORCE: THE EXTENDED INNOVATION TEAMS

Now let's talk about the aspects of the process, which, from our particular point of view, is closely related to the previous point.

As far as decision-making, a fundamental problem is that this process is seen as divine inspiration of the boss, where a single faculty of the great hierarchs is presumed and they are the ones who have the necessary criteria to take the decisions. Entitled by their studies in famous universities and sometimes in their experience, decision-making is seen and operated as individual prerogatives, and the problem is precisely this, that decisions are taken as an inspiration event of the boss and not as a process. A collaborative process by presenting arguments, facts, numbers, testing assumptions, and arguments in a cognitive debate, focused on making the best possible decision that the group in question can produce.

As Chip Sibony would say, "*if history gives us so many examples that people are routinely so bad at making decisions, how is it possible that the moon has been reached?*" The answer is that individuals did not do it, NASA did; ("*If people are so bad at making decisions, how did we make it to the moon?*"). It was not personal inspiration which succeded, but a group of people working collaboratively with a common goal; with a shared dream.

In spite of administrative practice where there are rigid organizational structures, there are also valuable human resources to improve the organization's capacity for innovation, but unfortunately, repressed and untapped skills.

Starting from the assumption that every human being has the creative potential, we can say that the main task of a manager of the 21st century is to take advantage of all the talent of the people. It is precisely the people who have this creative and innovative capacity who are the collaborators of the company; a first source of talent with virtually infinite possibilities of creating and imagining. In this sense, the manager in an innovation process should see himself as an architect of the process, not as the "*great inventor*" inspired by his knowledge and inspiration. (You should see yourself as the architect of the innovation process, not as a great innovator enhanced by your knowledge and personal inspiration).

In a decision-making process, more options are discussed, but assumptions and options will be tested by lowering bias or personal preconceptions and improving the quality of the process for future projects. To work collaboratively in an innovation process, one must have organizational maturity, or at least enough maturity for the group to work. If you do not have a minimum of maturity, the work sessions will be a real nightmare because groups tend to identify and "*defend*" their ideas affectively, since for a manager, it is usually very important to be right. In a process of innovation and decision-making, avoiding affective debates, where people "*win the discussion*", is not about "*winning*" but about building the best possible option with the different points of sight.

In a collaborative work to innovate is recommended multidisciplinary groups ideally, but in reality dear reader you have the group you have and with that you have to start, so that whether you have the dream team or the real team, you have to avoid attitudes or positions "*defense*" to defend views, but to actively listen to the construction and co-creation of ideas. That is, what develops is the capacity for active listening, not the ability to defend points of view or impose points of view based on rhetoric.

In this sense, the process of innovation is not based on organizational hierarchy, but rather on ideas, on the quality of arguments and assumptions, on a creative or co-creative discussion of ideas from a small group taking into account facts, and available information. In a process of innovation or complicated decision-making, testing the assumptions is a key aspect of the process, as the quality of these assumptions must be tested in a deliberate way, not only in a process of inquiry, to ask the right questions to the ideas exposed, but to test ideas with clients or potential users before investing money.

Apparently the process is slower, but in reality, it is not only faster but safer. This is what the authors call a cognitive debate, where several options are analyzed and the assumptions of these options are tested, analyzing facts, available information, or producing the necessary information. The center of the debate should focus on the same interest; on the same objective.

The objective is not only to debate and test the assumptions of the ideas made, but from different options, build co-create the best decision, design a relevant innovation that generates value. This type of process requires organizational maturity, maturity of the team, but above all, maturity of the leader that facilitates the process. The leader facilitating this process can make the big difference between guiding a successful or failing process. Organizational maturity is a (sometimes) complicated path, not immediately given, but it is worse not to start and the work of the leader is fundamental. The leader promotes a rational debate taking advantage of the different perspectives of the group, without canceling, minimizing, or ridiculizing options.

It guides the debate and experimentation with the client and the potential user, concentrating on the objective, project or problem to solve, not the hierarchy. If a company wants to have an organizational culture oriented on innovation, when innovation projects start, they have to concentrate on the quality of the process (the discussion), rather than the quality of the idea. With good processes in place that effectively promote the collaborative work, innovations that reduce costs or generate new products, services, or business models will come.

Let us recall the proposal of Peter Bregman in his article *Execution is a Problem, not a Strategy Problem* (Hbrorg, 2017).

The initial reflection invites us to think about the need for a focused and shared vision (Figure 4).

In his article, Bregman suggests 6 steps:

1. Define the Big Arrow (alignment of vision)
2. Identify the Highest-Impact People
3. Determine What They Should Focus On (Focus on Key Performance Indicators)
4. Hold Laser-Focused Coaching Sessions (focus-focused mentoring)
5. Collect and Share Data (collaborative / cross-cutting approach)
6. Amplify Performance

Integration of Multidisciplinary Innovation Teams

Once we have understood the importance of leadership in team integration and discussed the basics of collaborative work, we need to understand that team integration, as well as vision, character, temper, and maturity, also

Figure 4.
Source: Ortiz, S. & Moreno, T. (2017).

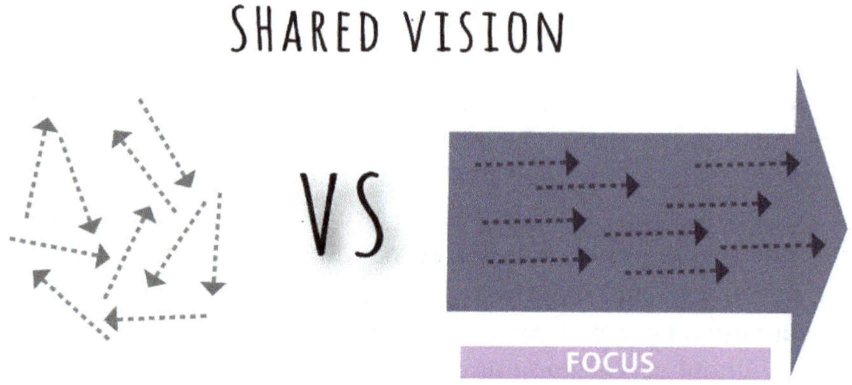

requires respect and confidence in the team. In Chapter 2, we talked about the effective communication model for high performance teams, which precisely dealt with these two previous elements used to reach a level of rapport that leads to loyalty.

However, it is necessary to remember that respect begins with understanding. Understanding and respecting the fact that we are all different, but together we are powerful, is not simple. It implies, as we said some rows above, a high level of maturity on behalf of the members.

In order to integrate innovation teams into organizations, it would be imperative to consider an interesting definition (among so many currently related to entrepreneurship theory) of the innovation profiles proposed by Tom Kelley & Jonathan Littman in the book *Ten Faces of Innovation.* (Kelley & Littman, 2005).

The authors suggested 10 different profiles that usually interact in these processes of corporate innovation:

- **First Group:** People who promote learning in organizations
    - **Anthropologists:** Those that provide the organization with new knowledge by observing human behavior and develop a broad understanding of the interaction between people and the products or services of an organization.
    - **Experimenters:** Those people who continually test new ideas, developing and validating prototypes under the framework of trial and error.

    - **Pollinators:** Those explore other industries, cultures, and business models and translate the findings into contributions that uniquely cover the needs of the business.
- **Second Group:** People who facilitate implementation and organization:
    - **Obstacle-Jumpers:** They are those people who understand that innovation is above the administration and know that their job is to overcome every obstacle that comes their way.
    - **Collaborators:** Are those who help to form solid teams and are usually among the group to offer multiple, solid and multidisciplinary solutions.
    - **Directors:** Not only contribute to forming a talented team, but also find the appropriate schemes to guide them under the principles of multiplier leadership.
- **Third Group:** People who imagine and design the experiences of value:
    - **The Architects of the Experience:** Those in charge of designing an experience full of delight for the client.
    - **Stage Designers:** Those who create a scenario where members of the innovation team can achieve their best creative performance.
    - **Those Committed to Care:** They focus on customer service beyond personal interaction.
    - **Storytellers:** People who build an internal moral and an external attraction from high-value narratives centered on the value of the human being.

The ideal would be to consider a profile of these in each of the innovation initiatives that are built within an organization. However, the starting point should be at least one of each of the three blocks suggested by the authors.

Evaluation of the Personnel and Strategic Adaptation to the Organizational Structure

When we talk about the integration of staff to innovation initiatives (I think this should be the process of recruiting staff in general), we talk about finding their level of motivation regarding the initiative; we talk about their attitude, commitment, and desire to bring to life a new product or service.

In several of the projects we have worked on, a very simple tool that has allowed us to be very agile in the initial evaluation of the people we include in the initiatives. Figure 5 is a graph that evaluates three elements.

Experience has made us understand the following:

1. The person does not know: result = gets trained
2. The person can not know: result = is adapted to the strategy according to their abilities.
3. The person does not want to know: result = is invited to find their personal motivation (inside or outside the organization).

When people do not or can not know, but they want with all their heart to participate in the innovation project, the best we can do is to provide them with the training and tools necessary for their development. Their personal motivation will eventually lead them to be apostles of initiative. This is the type of profile we are looking for to detonate innovation strategies.

The Follow-Up to the Integration of the Extended Innovation Team

The integration of human talent to the different innovation initiatives that are generated in the organizations is another of the great challenges to be met prior to the execution of the strategy.

Figure 5.
Source: Portillo, E. & Moreno, T. (2017).

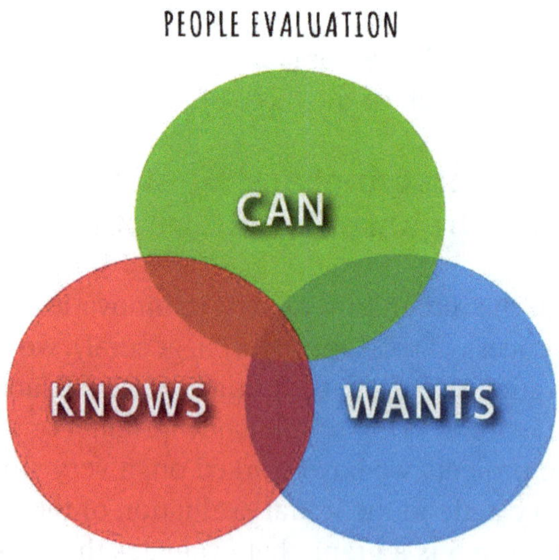

Figure 6.
Source: Portillo, E. & Moreno, T. (2017). Adapted from ZS Associates, 2006.

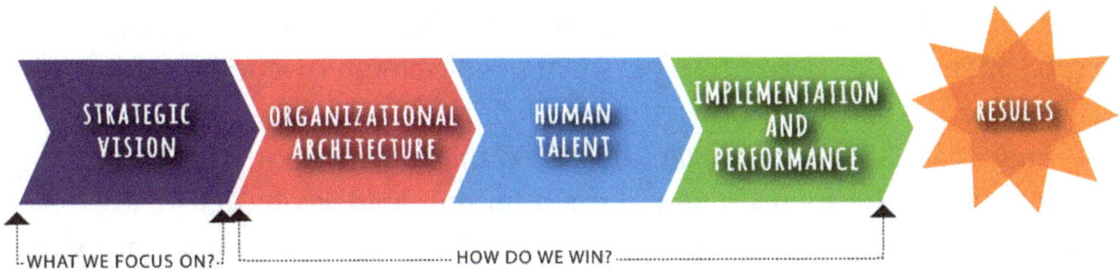

Once a vision has been defined, and the initiatives necessary to achieve that vision (and the organizational architecture to achieve that vision) are distinct, the next great challenge for the organization is the integration of human talent in accordance to their motivations and competences.

Teams of innovation and leadership in the face of change often experience greater wear than traditional teams. They face an additional factor: the resistance to change mainly by a myopia of comfort that generates resistance to change for lack of a strategic vision.

Figure 7.
Source: Moreno, T. (2017) Based on McKinsey & Company (2008) The GE-McKinsey nine-box matrix.

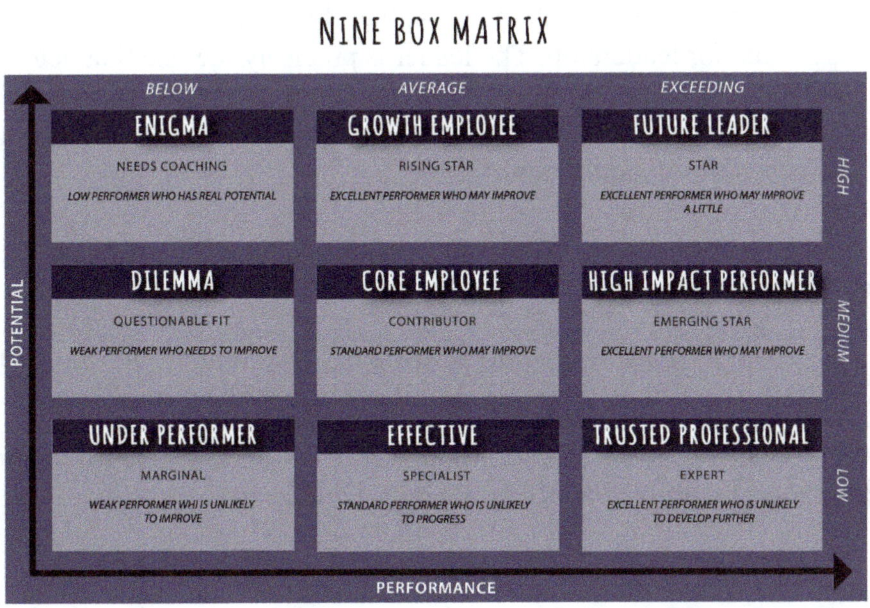

It is important to remember that every human being is doing his best from his own state of consciousness, which means that his maps and mental paradigms at that specific moment, indicate that the need for change does not present a stable scenario of personal development or does not allow him to visualize the scope or opportunities behind innovation. However, as we now understand, the human factor is also the most important asset of any institution; it is the focused and aligned human leadership that really gives names to institutions, so it is necessary to insist on the great opportunity that each of us has to potentialize their capabilities.

The human brain has 100 billion neurons, each neuron connected to 10 thousand other neurons. Sitting on your shoulders is the most complicated object in the known universe. (Michio Kaku, 2014)

As we have said before, human motivation generates an attitude, attitude generates an intention and intention generates a behavior. Reason why it is necessary to be measuring the levels of motivation and attitude of people who integrate innovation teams on a constant basis, to know how best we can support them in the process of transformation and change.

There are tools such as the GE-McKinsey 9 box model below (McKinsey, 2008), which helps evaluate the performance of our collaborators and their potential talent to detect emerging leaderships in organizations; although it is also useful to define strategies that help to enhance human talent in general.

The great challenge in applying tools like this is to find ways to provide a fertile ground for leadership. The leader is precisely the one that detonates processes of human development that encourage the maturation of other talents to the extent that the leadership is multiplied within the organization and for the benefit of a community.

REFERENCES

Hbrorg. (2017). *Harvard Business Review.* Available from: https://hbr.org/2017/01/execution-is-a-people-problem-not-a-strategy-problem

Kaku, M. (2014). *The Future of the Mind: The Scientific Quest to Understand, Enhance, and Empower the Mind* (1st ed.). Doubleday.

Kelley, T., & Littman, J. (2005). *The Ten Faces of Innovation: IDEO's Strategies for Beating the Devil's Advocate and Driving Creativity Throughout Your Organization* (1st ed.). Doubleday.

McKinsey. (2008). *McKinsey & Company.* Available from: https://www.mckinsey.com/business-functions/strategy-and-corporate-finance/our-insights/enduring-ideas-the-ge-and-mckinsey-nine-box-matrix

Toffler, A (1970). *Future Shock.* Academic Press.

Chapter 5
Prototyping

ABSTRACT

One of the most significant challenges faced by businessmen and entrepreneurs is to decipher which innovation or business ideas are worth developing and which ones they must discard. As it is generally recognized, the problem is not to have new ideas but to discover in which it is worth investing time and money. The recommended mechanism is to test the relevance and value of the concept with the user through a series of progressive, simple, fast, and cheap experiments. The instrument of experimentation par excellence is the prototype in its different manifestations, which allow discovering the relevant design for the market. This chapter explores prototyping.

INTRODUCTION

Make sure you are building the right "it", before you built it right (Savoia, 2011)

STEP 16. FROM PRETENDOTYPES, TO PRETOTYPES, AND PROTOTYPES

The innovative capacity of established and gestating companies depends, to a large extent, on the experimentation of new ideas and concepts related to the *product*, process, or business system. Due to its nature, the development of new concepts is subject to high doses of uncertainty, with failure rates ranging

DOI: 10.4018/978-1-5225-4023-6.ch005

Figure 1.
Source: Portillo, E. (2016), The Stragile Conscious Corporate Innovation Model.

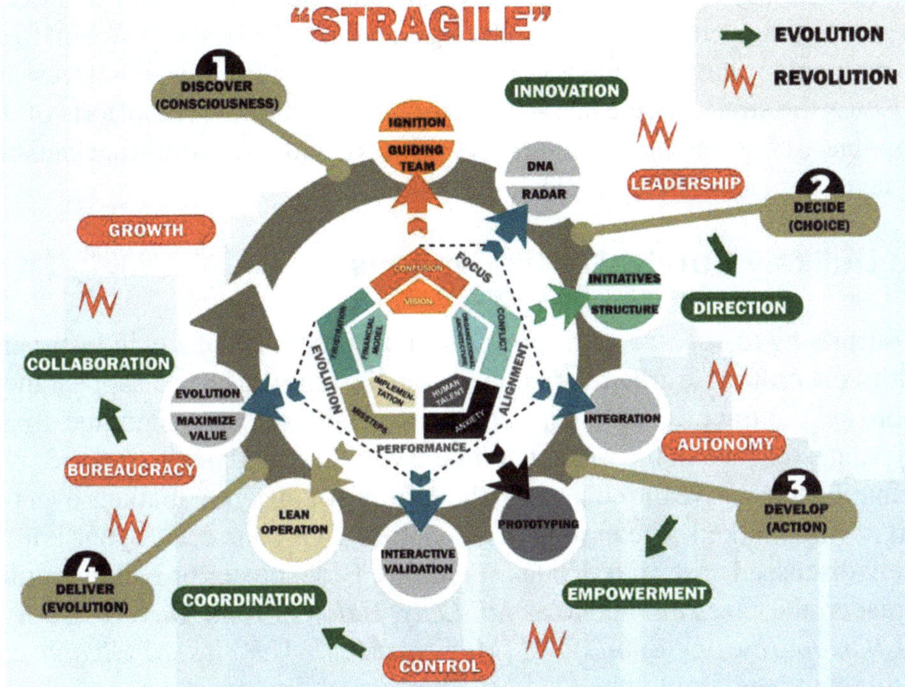

from 35 to 95 percent, depending on the category and source consulted. In the case of *gestating companies*, in Mexico 75% of these companies do not survive past the second year of operations (Gasca, 2014), depending on the validation of a single concept. Interestingly, the different studies on "*causes*" report reasons related to lack of income, poor planning, or implementation problems, but rarely mention the absence of the product-market adjustment.

Without the intervention of man, scientific studies have estimated the probability of reaching the adult life of a sea turtle—from the incubation of the egg—one in a thousand. This represents a "*failure*" rate of 99%. *Why should it be different for a concept in its attempt to develop a market?* The failure rate should not alarm us, a *sine qua non* of evolutionary systems. What is alarming is the lack of sufficient experimentation to discover and promote sufficient innovations, injecting vitality into the economic and social system in the process.

How can an established or emerging company develop new concepts without dying in the process? In the context of the agile strategy, the answer lies in the ability of the organization or the entrepreneurial team to quickly experience the assumptions of the strategy, seeking the best adjustment to an environment in changing and accelerating conditions. For this, it is necessary to discuss the strategic relevance of prototyping in the survival efforts of the companies, considering a competitive context with selection mechanisms that favor the emergence of relevant designs.

The Unlikely Survival of Innovations

It is surprising to observe in the population of companies of any industry, the healthy and profitable growth of some; and at the same time, the disappearance, fusion, or contraction of others. The vast majority is condemned to an irrelevant existence in the market. While trying to elucidate the reasons, it is fascinating to observe the changes in the ecology of different markets over the years, with a marked acceleration in recent times. Of course, everyone knows widely discussed cases like Kodak, Polaroid, Blockbuster or Nokia; but for the reader, the cases in Mexico as *Lili Ledy, Helados Yom, Aurrerá, Danesa 33, Chambourcy, Imevisión, VAM (Mexican Motor Vehicles) or Gigante* can not go unnoticed. And what about the high mortality among companies in gestation, despite the sagacity of the value proposition and the quality of the product in sufficient number of cases.

What explains the birth, exit, expansion, contraction, or extinction of companies in the markets? Although factors exogenous to the dynamics of the markets can partially explain it—economic crises, political and social instability, and conflict between nations—the causes here are endogenous. Among them, the most recurrent, i.e. the concept of innovation and competitive strategy, are insufficient to answer the question for theoretical or practical purposes. It is necessary an operative and actionable concept if some utility is intended in both cases.

The idea of a "*demography*" and "*population*" of organizations in a market (or "*ecology*" of institutions) suggests the existence of evolutionary processes in the variation of companies in different moments of an industry. Authors from economics and organizational studies have explored the evolutionary mechanisms of different industries, focusing on the ability of organizations to adapt to the environment. Accepting models between Lamarckism and Darwinism, they identify the existence of mechanisms of competition and selection, as well as the structures of variation, randomness, and reproduction. For this approach, the unit of analysis is the organization, concentrating on the capacity for change and adaptation, in the face of a constantly evolving environment. Despite the invaluable contribution to different fields of knowledge, particularly for the exercise of leadership, it does not answer the question: *what should the organization change to be able to adapt to the environment? Is it the structure, the administrative or operating systems? Organizational culture, knowledge or technology? The products, the distribution channels or the links of the value chain? All?* The evolution of markets and the need for adaptation are indisputable, especially when the process of change has accelerated. Then, *in what areas of the multiple activities should the company pay attention and how should those responsible for the results address it?*

Michael Porter offers a first approximation when establishing that the competition—fundamental characteristic of the evolutionary systems—happens between business units and not between companies (Porter, 1998). The business unit focuses on an offer and a market segment. In other words, it focuses on a product and a particular group of customers. Consequently, companies do not compete with each other, the products rival the choice of customers. The product-market binomial we will identify as "business formula", the right side of the business model in action. Established companies often have several business formulas; the emerging ones, only one.

This being the initial strategic premise, it follows that selectios—the central mechanism of evolutionary systems—is materialized through the decisions of the clients in their aggregate form.

A warning to the reader: competing products do not necessarily belong to the same category or use the same technologies. A fish fillet, a slice of pie, a commercial burger, a fruit, and any variety in the endless list of foods available, compete for the same space in the consumer's stomach. Less obvious is the competition between a traditional mousetrap and a high-frequency oscillator. In the same way a car, an Uber service, or an electric scooter do the same in the transport. The key to identifying competing products lies in the problem they solve.

If it is the product and not the company, what is the specific mechanism that conditions the survival and reproduction of these—and with them of the companies? Everett Rogers studied the processes of diffusion of innovations for several decades, identifying the attributes that favor the adoption of a concept, a technology or a product, being: (1) The relative advantage over other options, (2) compatibility with the customer's value system and consumption,

Figure 2.
Source: Beristain, L.M. & Moreno, T. (2017) The market selection mechanism.

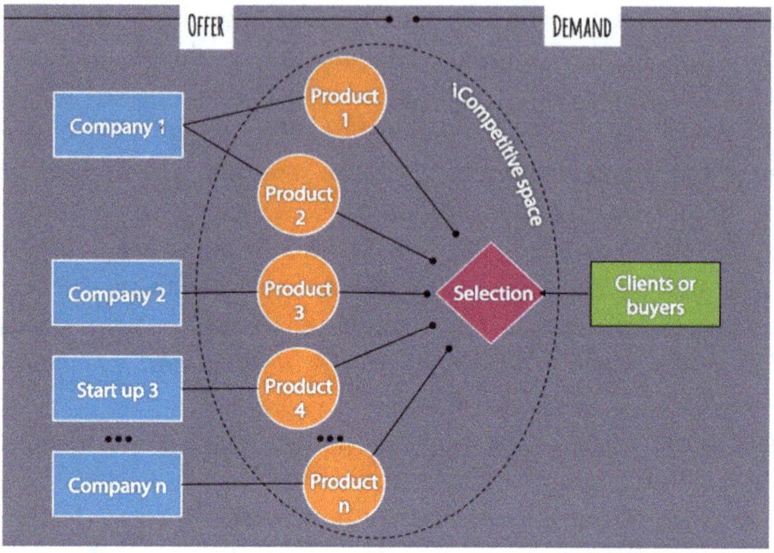

(3) the degree of complexity over the understanding and use of the product, (4) The possibility of experiencing it on a limited basis of low risk, and (5) the visibility of the product in the social system of the consumer.

Regardless of the stage of market evolution or the product or technology life cycle, products are exposed to constant competitive pressure to become or remain as a relevant solution as the "*relevant design*" in their category, a concept introduced by Utterback and Abernathy in 1978 (Abernathy & Utterback, 1978). Once the relevant design emerges, the evolutionary process between demand and supply sets up predictable paths of the market, process technology, value chain integration, and the relevance of the different companies linked to the industry. The opposite case occurs when the design does not reach dominant condition, condemning the company or the business unit, to extinction or marginal existence. This is true for the established company and/or the emerging company, regardless of the history of the first.

Of course, relevant designs evolve through small modifications or incremental changes in their form, meaning, and function; either by variations, mutations, recombination, and/or adaptations. In most cases, intentionally driven and subject to selection of the market (see following figure). This last one is the answer to the approach related to the mechanisms that condition the survival and reproduction of the products, and with that, of the companies.

Figure 3.
Source: Beristain, L.M. & Moreno, T. (2008) Evolution of the relevant design.

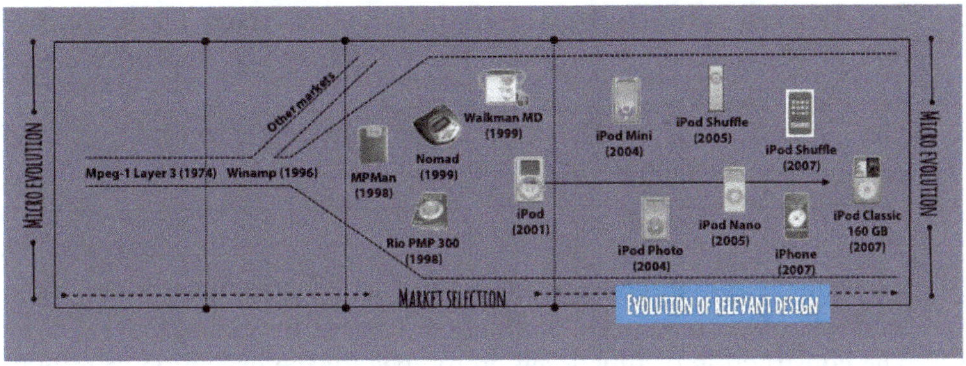

If the business strategy, the performance of firms, and the economy, as a whole, depend on the emergence of a relevant design, the question would be: *what is the design and how it improves the chances of success of a product in the selection process?*

The Leap From Idea to Reality

Starting from a simple approach, design is the process of transforming ideas into reality. All the effort of creativity to provoke the innovation of product, of process or business model, must be developed to introduce it to the market or to incorporate it to the set of tasks of a process, making use of the science, the technology, and the knowledge of the user as inputs for the exploration phase and discovery of the innovative concept. This stage is characterized by market uncertainty, the lack of definition of the "*production*" objectives and the proliferation of specifications, from which the need to experiment soon and repeatedly with the object of receiving feedback and generating knowledge. This effort demands multiple experiments; all with a single objective: Learn about the necessary characteristics of the product to incorporate them into the specifications, trying to satisfy the five attributes proposed by Rogers.

1. The overall design process, with different degrees of complexity, cost, and time remains the same regardless of whether it is the development of a medical device, car, electronic device, app, or translucent-conductive concrete: exploration and discovery, on the user's problems—articulate,

disjointed, dissatisfied and even unknown—or frustrations in the performance of a technology.

2. The ideation or identification of concepts based on scientific, technological, and user knowledge validating the acceptance and development relevance.
3. The construction of models and prototypes, physical and virtual, from the previous understanding, with the purpose of conducting experiments and proof of concept of the technology, such as acceptance and desirability of the user.
4. Experimentation of concept and proper acceptance, incorporating the feedback of the results to the design criteria of the concept.
5. The construction of functional prototypes for the execution of tests in real conditions of use or through simulators, to verify the fit between the design, the regulations, and the performance.
6. Finally, the integration of learning to fit the design, returning to the original approach in an iterative process of refinement, until the introduction of the concept to the work of engineering for manufacturing, the market, or the routines of the organization.

At first glance, it is difficult to differentiate the design process from science or engineering, and sometimes from the arts. In a sense, design understands and is understood in all disciplines and forms of knowledge. Let us reflect for a moment on all the objects, means, and tools arranged in the room in which the reader is at the moment, including the electronic device for reading this document in digital format or the paper volume of the physical version. We simply live in a world designed voluntarily, intentionally, and with a clear purpose by men. To put things in perspective, about 353.7 million individual products are available through Amazon, excluding books, media, wines, and services; none of them resulting from an accident, spontaneous generation, or the will of a higher intelligence.

Each "*thing*" is adequate and practical, thanks to a design process. Depending on the knowledge and technology involved in the "*thing*", the process acquires idiosyncrasy. Take as an example pharmaceuticals and the development of mobile applications, to illustrate the differences.

The Design of Pharmaceutical Products

Unlike other activities of inventiveness and technological development, the life sciences innovation is characterized by conducting tests in humans and being

Figure 4.
Source: Beristain, L.M. & Moreno, T. (2017) General design process.

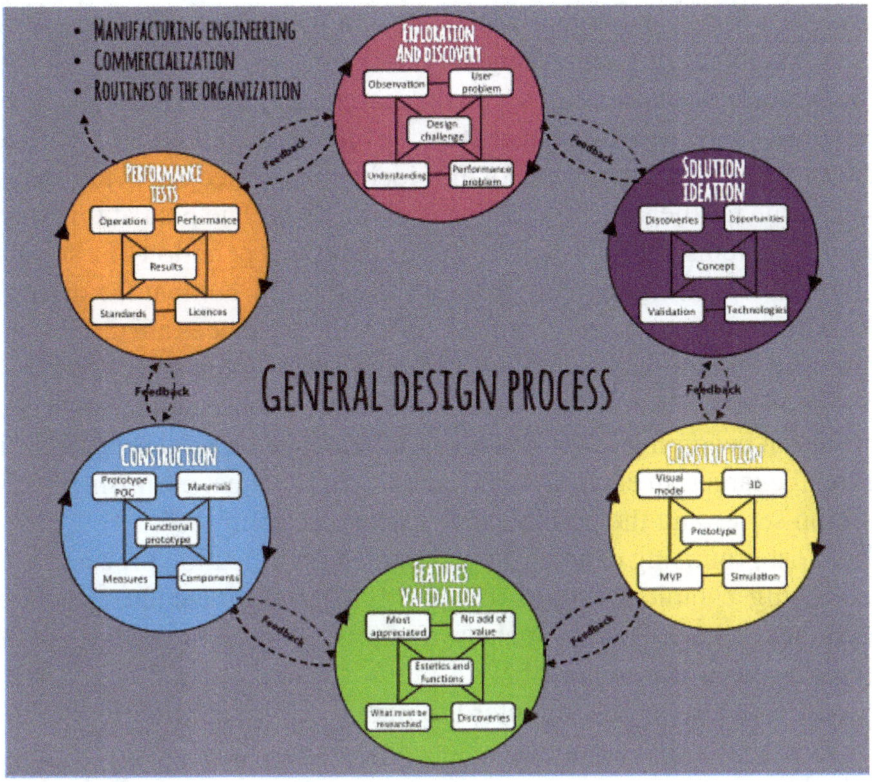

subjects to a vigorous regulatory regime throughout the chain of development and commercialization. The system of choice and consumption differs clearly from any other market: the end user barely intervenes in the decision and has partial information, which falls on the treating doctor and sometimes on the availability choices established by the hospital administration or the public health system. It is not clear who the customer is, traditionally defined as: *The agent that covers the cost of the product*; sometimes being the end user, the family, the public sector, or insurance companies. The sector includes a broad spectrum of technologies such as medicines—whether small molecules or biotechnological—medical devices, diagnostic systems, functional foods, and possible combinations between them. Due to their nature, they usually have applications or possibilities in the global market and it is necessary to consider, in the technology evaluation, the feasibility of protecting, conducting clinical studies, and complying with accepted and different regulations, laws, and standards in each of the target markets.

Pharmaceutical innovation can be conceptualized in three major moments: *basic research and development, clinical research, and commercial introduction.* Figure 4 summarizes the design process of this industry prior to the escalation of manufacturing and marketing.

The exploration phase begins with the search, discovery, and identification of proteins— which may be receptors or enzymes, linked to a particular condition. These are responsible for the biochemical processes in the body and consequently, for the structural changes in the cells. The goal at this stage is to identify proteins with the potential to be the target of action of a substance (drug) to inhibit or improve its function, and thereby combat the disease.

Once the target has been identified, different substances are tested—large pharmaceutical companies can have libraries with more than 2 million of them—in order to identify the main candidates (lead candidates) with the possibility of becoming a new active ingredient when it produces the desired effect on the target. In each case, the potency of the most promising is tested in dilution series and the possible side effects are examined.

The most promising candidates undergo an optimization process, adding or decreasing chemical groups, or modifying the molecular structure in order to meet other criteria—such as affecting only the target and not other molecules, not to decompose before producing the desired effect, and be soluble in water to be able to enter the organism (in addition to producing the expected effect). The optimization process can take several years and is applied by a patent once you reach a prototype with the desired criteria.

This stage concludes with pre-clinical studies to determine if the molecule is capable of treating the condition (pharmacodynamics) and to study the absorption, distribution, metabolism, and excretion of the drug (pharmacokinetics); knowledge necessary to determine the form of administration. These tests are initially performed in computer simulators—in silicon, to pass tests on cells, tissues or bacteria—in vitro, and finally the study in animal models (in vivo, which are regulated by strict procedures and protocols of investigation).

When the results of the preclinical tests are completely satisfactory and authorization procedures established by the health authority have been covered, it is possible to perform tests on humans, thus initiating clinical research.

In a first stage or Phase I of the study, the active ingredient is administered to a small group, usually 7 to 30 healthy volunteers to test for safety, tolerance, and other effects on the body. The protocol must be authorized by the health authority, including the informed consent of the individual and the monitoring

Figure 5.
Source: Beristain, L.M. & Moreno, T. (2017) Pharmaceutical design process.

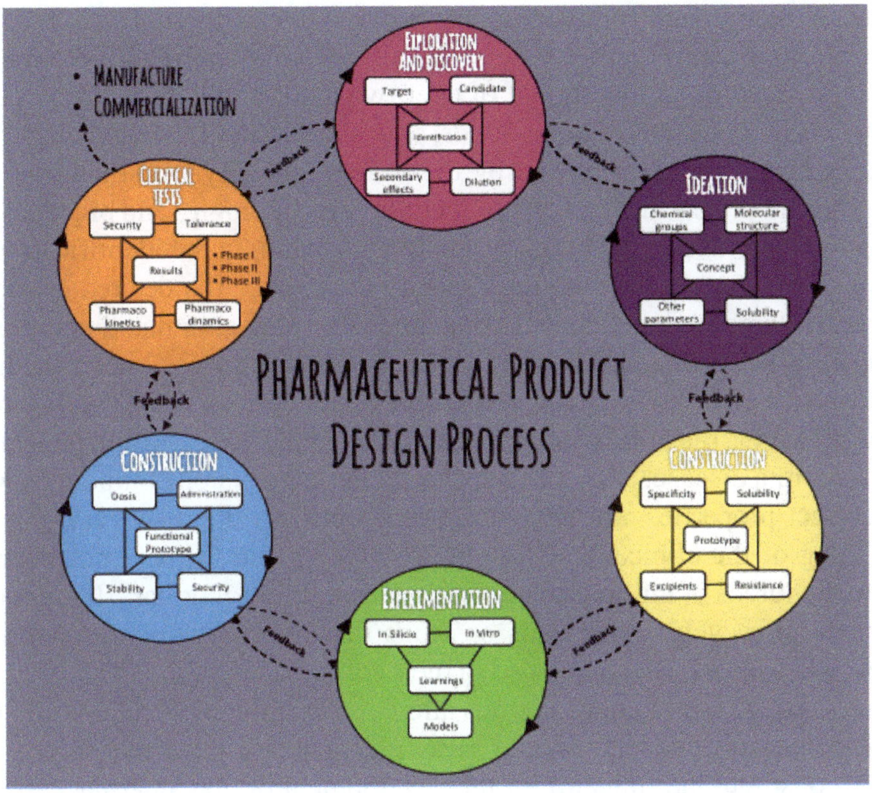

of an independent ethics committee. When the drug proves to produce no serious side effects or toxicity, Phase II will proceed.

In Phase II, the compound is administered to a group of individuals—from 100 to 500 subjects—with the condition to be treated, in order to determine the safety and efficacy of the drug candidate. The study is conducted by doctors and hospitals in different regions of the world. New pharmacodynamic and pharmacokinetic tests are performed. The efficacy and benefits of existing treatments in issues such as safety and convenience are studied, being necessary to show a clear advantage with the risk of not being approved by the sanitary authority for its commercialization.

If the results were entirely satisfactory, the tests are scaled to Phase III for groups of patients between 500 and 20,000 subjects to be treated, seeking to confirm the effectiveness, dose, safety and possible side effects. Between 34 and 37% of the drugs that started Phase III studies will be discarded. Those drugs with positive results, approximately 12 out of 100 that initiated the clinical studies will be submitted to the approval of the sanitary authority for its production and commercialization.

From the discovery of a potential drug until its approval, between 10 and 15 years with an investment of over two trillion dollars is invested. Of every tens of thousands of compounds, only six to eight will conduct clinical trials and one will be approved by the health authority. To streamline the development of pharmaceutical products, new technologies and practices are being implemented in the industry. On the one hand, in the identification of active substances (lead) with the capacity to produce the desired effect on the target, large pharmaceutical companies use new automated equipment to perform more than 1,500 simultaneous tests (a process known as high Performance or HTS), reducing the identification time of months or years to a few weeks. On the other hand, the knowledge of the structure and molecular pathways of a disease, allows to select with more precision compounds with the correct properties and reduce the time between the discovery and the beginning of the clinical tests to only four years and the investment from several hundreds of millions to hundreds of thousands of dollars. It is enough to renew or recycle compounds or drugs that have not fully shown their efficacy and have been abandoned in their commercialization to use them as a source for the treatment of various ailments once the molecular structures are known with precision. While these practices are relatively recent and do not constitute the standard for the development of new pharmaceuticals, in the near future, they must modify the relationships between the different actors in the innovation chain of the industry.

The Design of Mobile Applications

We are witnessing the spectacular and growing importance of the "*economy of apps*" (or the economy of mobile applications), becoming the main source of disruption in the various spheres of human activity. Mobile devices, along with apps, have freed us from desktop computers and laptops that enabled use of the Internet. Together with the Internet of Things (IoT), it promises the omnipresence of the will, the creativity and the capacity to consume and produce information through a device, approaching the incarnation of a cyborg in a symbiotic relationship with technology, as proposed by Manfred E. Clynes and Joseph CR Licklider, respectively, in 1960.

The dynamics of the sector allows observing the evolutionary processes of gestation, selection, reproduction, mutation, or extinction in a short period of time. As of 2008, the date of the introduction of the App Store in iOs and Android Market (renamed Google Play in 2012), the offer has gone approximately from 60 thousand to just over 5.7 million apps in mid-2016, with the App Store and Google Play concentrating 74% of the distribution (Statistacom, 2017). The value of app sales reached $ 35 billion in 2016; of these, 69% were captured through the Apple Store. The gaming category is the main billing engine, accounting for 75% of revenue for the Apple Store and 90%for Google Play. If you look at advertising revenue, the value of the apps' economy is $89 billion, growing at double-digit rates. Despite the growing supply, 70% of the use is concentrated in 200 apps and about 80% of browsing time in three of them. For active users, the most popular are Facebook, Messenger, YouTube, Google Maps and WhatsApp. In the games category, the relevant designs in 2016 were Pokémon Go, Candy Crush, Clash of Clans, Subway Surfers and Clash Royale (Appanniecom, 2017).

The design and development of apps can be conceptualized in three great moments: (1) The one of exploration and discovery on the problems and the opportunities of the user, (2) the ideation and development of the applicatio, satisfying the associated tests, and finally, (3) the launch in the relevant distribution platform. Figure 5. summarizes the pre-launch application design process.

The exploration and discovery phase starts by identifying the "*problem*" of the potential user. Resolved by discrepant means, frustrations go unnoticed or underestimated for usual reasons. Before WhatsApp there was "*efficient*" media, just as before Google Maps, a Roji Guide was sufficient. *What about apps like Uber or Airbnb, whose initial proposal was rejected by experienced investors due to insubstantiality in the light of custom?* Occasionally, the user is different from the client, being necessary to formulate solutions with differentiated value proposition. Consider the case of Uber. The user may be a young person with the need to satisfy the demands of transportation for social reasons, but the client, usually the parents, wants the peace of mind to know who is moving him, by what route, and at what time.

At this stage it is necessary to go out, observe, ask and, if possible, experience the context first hand to understand the emotions, meaning, and relevance of current practice. Simultaneously, explore the possibilities of improving the experience, solve the problem, or completely recreate the uses and customs, using resources for mobile devices.

Once the problem is defined, the solution is devised, incorporating the learning about the user in the previous phase. Of course, it is common to find solutions in search of a problem; however, when the selection of the relevant design depends on the choice of the user, it makes sense to incorporate it into the configuration process from the beginning. At this stage, drawings and sketches are developed to conceptualize the interface, operation, and characteristics of the application. The fundamental thing about the draft is to be able to present the concept to people who clearly satisfy the profile of the user and the client; to confirm whether or not it is worthwhile to continue with the project. The objective is not to sell the idea or to the one who proposes it, but to discover information relevant to the winning concept (a first selection).

Once the concept is validated, a digital mockup is created with the possibility of simulating user interaction, composition, and operation of the application. In this phase, it is essential to define information architecture, contents, and

Figure 6.
Source: Beristain, L.M. & Moreno, T. (2017) Mobile Application Design Process.

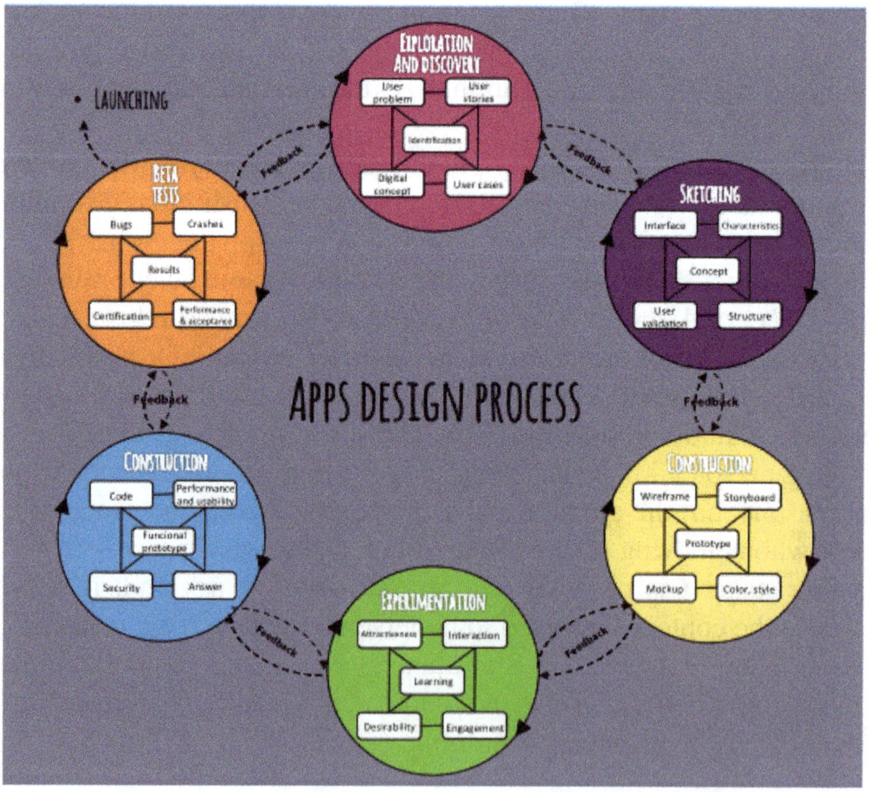

interface in detail: the type of information, available functions, navigation rules, usage scenarios, etc. Similarly, logos, images, colors, fonts, style, and any element that give "*personality*" to the application are defined. Each of the design decisions should be properly documented for the development phase, including the storyboard as the roadmap on the relationship and logic between the different screens of the app. This documentation is the basic input for the development stage. In this phase, the design of the service platform is decided, which includes servers, databases, and data storage solutions.

With a digital prototype capable of simulating the operation of the application, we proceed to design an experiment to test the feasibility of the concept. This means making the proposal known to potential users by various means, obtaining information about the experience, desirability and price of the app to validate the assumptions of the application—a second controlled test of selection. It is also a vehicle to refine the solution and perfect it. At this stage, you must choose the distribution strategy: App Store, Google Play, or both. You can also consider other options like Microsoft Store or BlackBerry World.

If the results of the experiment justify it, proceed to the development of the application itself. It is in the development phase that the codes are deployed using different languages and publishers like XCode and Android Studio, through successive series of debugging. Simultaneously, the visual language of the interface or user interaction screens is created, considering factors such as typography, spaces, colors, images, animations and transitions, emphasizing user actions discovered in the previous phases. Fragmentation to visually adjust to different devices in terms of screen size, DPI's and aspect radios, must be resolved to ensure addressing the target market. During the process, it comes from an Alpha version to verify the stability, errors, and operation, to a Beta version to be tested under real conditions of use.

The Beta experimentation aims to review the operation of the application with real users to validate the technical performance as the acceptance of the user. Ideally, it is tested in different environments, conditions, and devices to ensure the absence of bugs and crashes in its operation (the stability of the system is one of the main satisfaction criteria at this stage). From the market perspective, the superior performance over other solutions and the acceptance of customers are analyzed in detail. Ideally, customers should express clear enthusiasm for the app and demonstrate an "*addictive*" commitment to its use—a third user-selection test.

Once the problems observed during the beta test are resolved, it is possible to apply to publish the solution on one or more of the distribution platforms.

Some, like the App Store, have a pre-distribution approval process, through reliability, security, and content analysis. There is no reliable statistic on the total reject rate in the App Store, but by comments from programmers it can be significantly high. In the case of Google Play, there are fewer restrictions for publication; however, those of poor quality or download are discharged quarterly.

There is no published information on the success and survival rate of mobile applications; however, it is estimated that less than 0.01% of consumer apps are financially successful for their developers, with about 95% downloads in free applications and 90% below the 500 daily downloads line. With more than two hundred platforms for the development of apps and the explosive growth of the offer, the expectations of the clients have risen to be willing to pay for a solution of this nature. However, the possibility that a person programming in the room of his house or apartment can change the world, remains latent.

Experimenting With New Concepts

How many new concepts OXXO (supermarket stores) tests annually to evaluate the possibility of introducing them to the regular line of products of the small format store chain? (There is no "*official*" data, but by talks with company executives, the figure ranges from 160 to 400 products per year. In the opinion of the author, this is the difference between an established company and a gestating company: the first, has the possibility of experiencing multiple concepts, while the second, only one). Contrary to the popular knowledge of business specialists, the growth of the company is not the result of a mere "expansion strategy" or a new model of partnership for the management of all aspects related to the point of sale. The expansion or reproduction is the result of a successful design of the value offer. As we have previously stated, when a concept does not reach the category of relevant design, it tends to become extinct or marginal. But even relevant designs, in their vast majority, have a validity and an expiration date—in correspondence with the widespread concept of life cycle, whether of the product or the technology. When a mature product begins to decline, in a short time it is replaced by a new one with growth potential, and if it does not capture the customer's attention, they substitute it for another one, successively, until finding the offer desired by the consumer (Alcocer & Campos, 2014). The same happens with the location of the point of sale: If in a period of six months it does not reach the projected volume, they close the operation and are installed in a new location, sometimes, on the opposite side of the sidewalk. More than 14 thousand points of sale expand the possibilities of experimentation of new concepts. As an example, at the end of 2016 they began the pilot test for the distribution of products acquired through Amazon, with the installation of lockers at eight points of sale (OXXOInmuebles, 2017). How many of the more than 14 thousand points (FEMSA, 2012) will have this service available? That response, only the market and the results of the tests can answer. However, the company knows that a quarter of its annual profits come from new products and can not afford to validate by mere intuition or perception of business. For this reason, it has a systematic procedure to capture, analyze, and test new concepts through a planning office to evaluate and take advantage of revitalizing business opportunities.

The strategy of exploration, discovery, experimentation, and validation, is not necessarily solve with the "*departmentalization*" of the creative function in the company. Consider the case of Cinepolis, one of the most innovative companies in designing memorable entertainment experiences. When the threat of digital and online formats portends the survival of the projection theaters in the world, Cinepolis has consolidated in three continents with 4,910 rooms, 870,000 seats and 312 million attendees; placing it as the second major chain worldwide by the number of spectators. (CinepolisCorp, 2017) To ignite the "*innoduct*" of concepts, in 2011 the Cinepolis SeedCamp was created. A program of acceleration of projects outside the company that improves the experience in the cinemas. By 2015, it has switched to Cinépolis Accelerator, receiving more than 500 projects worldwide by call and expanding support with advice and corporate venture capital, to concepts not related to the core business of the company. From each call, 7 to 9 projects are selected to participate in the 19-week program (CinepolisCorp, 2017). This way, Cinepolis takes advantage of one of the modalities of open innovation, to discover those concepts that favor the strengthening of the relevant design in the industry.

Multiple strategic arrangements to discover and develop the design of new business formulas arise in a continuum between the own research and development centers, and open sessions using Sprint (Designsprint, 2017) or Design Thinking (DesignThinking, 2017) methodologies. Depending on the secrecy of intellectual property or the complexity of technology, even coordination with various actors in the relevant development ecosystem is encouraged—including cooperation between competing firms where there is no "*conflict of interest*" in the competitive space in question. What is relevant in all possibilities of development is the focus on experimentation.

Being entrepreneurship and business management a practical and prudential knowledge, decisions must be a product of evidence and data, not "*business intuition*" or perception. Even assuming a conscious objective and rational position, the human being is subject to making decisions from inappropriate or incomplete information—under conditions of "*limited rationality*" as Herbert Simon referred to it in his 1945 work. Experimental science is essential and necessary as it tests any idea, assumption, or paradigm with an opening that allows to observe the technology, the environment, and the users in their real dimension. The ideas or beliefs of the entrepreneur, some investors, the expert consultant, or the business professor should be irrelevant to verify the validity of a concept. Science is subject to multiple errors and inaccurate conclusions, but the results are always declared with the precaution of "*tentative*" and in the form of a hypothesis that is feasible to be refuted by new data, that is, falsifiable. In the same way for business and technology, it is the "*facts*" that speak, not the business strategist or the expert in knowledge. The success of science lies in its method: *one intended to correct errors or to incorporate the effects of a new environment; to question "obvious" truths or common sense...For obvious, at some point the earth was "flat."*

To use the scientific method or, rather, the "*scientific thought*" to the tasks of the company and of the entrepreneur, like to the development of the inventions, is not new. Without going into too much depth, Adam Smith refers to the decision to choose a commercial activity throughout his work of 1776, as an "*experiment*":

The landowner can afford to attempt experiments, and he is generally willing to do so. Failed experiments cause a moderate loss for him. The successes contribute to the improvement and superior culture of the whole country. (Smith, 2000)

Throughout his work, he discusses the conditions under which the experiment must be carried out by the government and in which the risk must be assumed by individuals.

In his 1880 treatise, he elaborates his thesis starting from "*scientific thinking*", even devoting much of the first chapter to differentiate the descriptive sciences from the experimental sciences:

I have said that the farmer, the manufacturer; the merchant, carry out their business to make profits the knowledge previously acquired, applying them to the satisfaction of human needs. I must add that they need another kind of knowledge, which can only be obtained through the practice of their respective occupations, and must be called their technical ability...In the arts [referring to activities of a technical nature] there is a certain kind of perfection which results only from repeated tests, sometimes successful, and sometimes on the contrary. So, science alone is not sufficient to ensure progress, without the aid of the experiment, which is assumed to be more or less risky, and does not always compensate the adventurer whose gain, even if successful, is moderated by competition. (Jean-baptiste, 2001)

Like Smith and Say, other authors relevant to business and entrepreneurship have drawn attention to evidence-based management. Particularly Peter Drucker in his work of 1986 (Ucker, 1988), Stefan Thomke in 2003 (Tomke, 2003), and Eric Ries in 2012 (Ries, 2011). From the latter, the need to put every concept to succeed in innovation and entrepreneurship has become popular, with the emergence of multiple methods, most of them with little seriousness and substance but in the right approach.

Prototyping

How to proceed with experimentation to discover or validate new concepts? Science and technology, two different acquaintances, have established a symbiotic relationship to progress: *knowledge about the functioning of nature allows the development of new and sophisticated observation and measurement technologies (instruments), expanding the possibilities of discovery of science in a successive process of mutual reinforcement.* Similarly, the established company and the entrepreneurial team currently have multiple tools to perform experiments quickly, easily, and at low cost, significantly reducing the natural uncertainty in projects of this nature and increasing the chances of discovering a relevant design. The specific *"instrument"* for experimentation oriented to innovation and entrepreneurship in business is the prototype. Defined in a first approximation as the modeling of the idea, the concept, or the opportunity, regardless of whether it is a complex technology or a physical "*thing*" subject to the laws of Newton, a computer program or a service, for the sole purpose of *testing acceptance, use, design, and operation.* In general, any means to "*materialize*" an idea or concept; creating unique references for those involved—thus avoiding errors of interpretation and understanding—is considered a prototype.

In the proposed model for the design process, there are at least three moments in which the construction of a prototype is necessary: in the stage of ideation, in the acceptance test of the value proposition, and in the verification of operation. For each of them, with their respective iterations, a different type of prototype is required, from a simple sketch on paper, to a finished model in all its elements for the reproduction and scaling engineering. The total number of prototypes to be built may vary depending on the results obtained in each experiment and the technologies considered in the development of the solution. As reported by Dyer and Martin, Thomas Alva Edison had to perform more than nine thousand experiments and an equal number of prototypes to solve the problem of a battery (not the electric bulb, as popular knowledge asserts). The clarity in the objective for the construction of the prototype conditions the degree of accuracy or fidelity in the representation of the final product.

Considering the objective variables—from the validation of an idea to product certification through aesthetics and performance tests—and fidelity—from a pencil drawing to the final product—it facilitates the understanding of the multiple forms a prototype can assume. Because of the history and culture of the different professional "*guilds*", each discipline has developed its own jargon to refer to different types of modeling, causing unnecessary complexity for the novice in the subject. However, the objective-fidelity relationship allows the prototyping strategy to be activated.

Prototyping From the Objective-Fidelity Relationship

One of the most common mistakes in the search for the relevant design is to concentrate efforts on the execution of the idea. However, the real goal of all prototyping work is to learn quickly and inexpensively, to be able to implement faster. Thus, the process must go in an upward direction, resolving the learning objectives in order of priority. These objectives are but the hypotheses and assumptions behind the idea, starting from the question, *is it worth developing?*

If the intention is to develop a relevant design in the market, then the first learning refers to the understanding of the problem. It does not matter whether the solution resorts to science—such as biotechnology and biomedical—or whether it deals with consumer products—such as mobile devices or digital applications—the starting point is field and documentary observation. There is no prototype. Instead, there is a documentation on the nature of the problem, or as many prefer to call it, grief or frustration. What is interesting in the exploration and discovery phase is to find evidence of the existence of the problem, from the perspective of the client: *What particular forms of solution are currently used? How much are you willing to pay to solve the problem? How many have the problem?* By reflecting each of these questions, translated as hypotheses, it is possible to use the sufficient information available on the Internet, identifying reliable sources and using certain "tricks" to derive the relevant data when it does not exist expressly. Ideally, each of the existing solutions should be evaluated and unresolved issues recognized. The greater number of solutions, regardless of their technological nature, the better: It is a sign of the absence of a relevant design.

Consider nuclear magnetic resonance technology. All doctors and physicists know that by applying heat it is possible to destroy cells, useful knowledge when thinking about cancer and tumors. However, current equipment can not be manipulated with degrees of precision and specificity necessary to avoid damaging healthy cells. For those who have a family member or friend undergoing aggressive therapy of this nature, they clearly understand what the problem is: The side effects of treatment. In this case, the research objective is clearly defined, focused on the manipulation of the emissions of magnetic resonance equipment to apply selective and specialized radiation.

However, in most innovation projects, the problem is not obvious. Towards the end of 1994, Iusacell dominated the Mexican mobile telephone market. In order to deal with the economic crisis, he chose to focus on the business segment and high purchasing power, while Telcel, a modest competitor, decided to introduce the concept of prepaid service as a mechanism to enter a new segment: Youth. Originally, the mobile telephone was aimed at solving a communication problem for business executives or high officials in the public sector. However, the prepaid system made available to adolescents a socializing tool promoted by parents' desire for security. Specifically, the problem is the parents (the client) and not the young (the user). Such was the urgency of the problem that Telcel not only developed a dominant position but did so in the years of the most acute economic crisis experienced by modern Mexico.

In the task of understanding the problem, the innovator visualizes different concepts with the potential to solve it. These ideas, to be developed, require heavy doses of investment in time and resources of another nature. *Is it worth doing?* To answer the question, it is necessary to validate at least two hypotheses: (1) The existence of the problem, partially resolved in the previous stage, and (2) the desirability of the proposed solution, always from the client's perspective or from who decides the outcome. This requires a first experimentation and for this, a representation of the concept in the form of a dirty prototype or low fidelity. Other names have received these representations, like *"mule"* in the case of software or pre-totipo for physical products. A drawing in paper, a construction in cardboard or the use of images in PowerPoint, is enough. The innovator should go out to ask, observe and learn. In principle, the target audience should satisfy the profile of the customer hypothesis—the avatar or "*person*" as reported by some Apps developers or professional designers—to receive first-hand feedback. In the process, the hypotheses of the problem, the solution, and the client's own profile are validated. In any case, there is always the possibility to reformulate any of them (pivot in the language of *"lean thinking")* and iterate, until all three are validated.

Sometimes the idea needs to solve a fourth hypothesis: The operation of technology. In these cases, the initial prototype acquires the form of a mathematical modeling but also in the construction of a prototype for proof of concept. In 2004, J.B. Straubel and Gene Berdichevsky, of the first employees of Tesla Motors, had recognized the potential of lithium batteries used in computer equipment to be used in electric cars. To prove their potential, they built several battery packs in the first garage, as prototypes to demonstrate the feasibility of technology, before thinking about the design of an electric car (Vance, 2015). The support of evidence and previous studies are necessary for each of the technologies or components of the final system.

Once these previous hypotheses have been dealt, the following group of objectives is related to the validation and consideration of the form, functions, and characteristics of the solution; thus exploring the aesthetics, user interaction, and general product experience. The construction of digital models or representations is necessary, facilitating the interaction of the potential customer with the proposal through the simulation of product performance. The appearance of a finished product is desirable to capture or build the relative advantage over existing solutions and compatibility with the customer and user value system. Programs for the development of "*mock-ups*" in the case of digital products; the construction of models in AutoCAD and 3D printing; the use of virtual reality or the use of conventional materials such as wood, plastic, or foam for physical products including those called "*services*" which allow to transit in the experimentation. Together with the system design hypothesis, it is a propitious moment to verify assumptions such as cost and sales price; to validate the economic feasibility of the project.

Resolved the design in its form and function, it is necessary to validate if the technology behaves as it is presumed using the construction of functional prototypes or proof of concept (POC). If the technological proposal is disruptive, it is advisable to proceed with this construction before or in parallel, to the solution of form and functionality previously described. In most cases, the prototype takes the form of an authentic *"Frankenstein"*: a cable here, an energy source there, a circuit board on the other hand and a computer interface in low resolution. In the case of a pharmaceutical product, for example, it is a laboratory test (in vitro). For software and applications, an evolutionary development with the possibility of adjusting to different operating environments and concentrating on those blocks properly understood, favors the construction by modules of the system. Ingenuity and creativity are fundamental to proof-of-concept: When Tesla's first prototyping team developed a functional battery pack, it was installed in the chassis of a Lotus Elise to test the cooling mechanisms through air and liquid. This was the basis for the company's first vehicle: The Roadster (Vance, 2015).

Once the operation of the parts and the system, as a whole, is resolved, it is necessary to comply with industry regulations, incorporating them into the design decisions. To a greater or lesser extent, every product is subject to norms and standards. To satisfy them, prototypes are built close to the final version in order to subject them to performance tests such as safety, resistance, behavior to extreme conditions, effect on the system and its parts, and an endless number of possible requirements depending on the product and of the industry. This is equivalent to pre-clinical and clinical trials of a drug; Safety tests, driving, lifespan of parts, fuel consumption and collision in a car; soil studies, structural analysis and engineering for a building; or tests of efficacy and safety for a functional food. The tests and their results may relate to the parts or to the total system of the product, including materials and components. This is an experimentation of a technical nature largely omitted from slender thought and design thinking, whose neglect can produce enormous losses for the enterprise or the entrepreneurial team, as evidenced by the cases of the Samsung Galaxy Note 7 (Forbes, 2016), Hoverboards (Montgomery, 2016), or *"Shiftergate"* in more than one million units of Fiat Chrysler for *"flaws"* in the gearshift, whose operation was expected, but with an unconventional design (Gorzelany, 2016).

Finally, the definitive model or commercial prototype is built as a standard for construction, scaling, or mass production. This includes the integration of documentation, plans, engineering, and specifications needed for certification and reproduction. Generally, a pilot test or a limited production is performed to experience the performance of the product in a controlled but actual market or environment. Even once introduced for regular adoption, it remains subject to behavioral and performance verification by competent authorities or institutions, such as the pharmacovigilance protocols administered by

COFEPRIS, the Vehicle Safety Verification Standards of the Secretariat Of Economy or the laboratory studies realized by the Prophet, to mention some. When the product has a global reach, knowledge of regulations in other national markets must be incorporated throughout the development process. Prior to the launch or introduction, other design-related issues such as production, distribution, packaging, operation guides, engineering and user manuals, the landing page and disposal of consumption, including technical support and customer support teams must be adressed.

Figure 7 summarizes the objective-fidelity relationship, offering an approximation to the taxonomy of the prototypes. The curve represents commitment and investment in time and cost, as it is carried through iterations, from one objective to another and in the approach to the final prototype. Taxonomy applies conceptually to any discipline; however, each one has developed its own. In architecture, low-fidelity prototypes are called sketches and drawings. For the definition of the characteristics and to attend to the needs of the user, seed plans, space distribution and perspectives are used, supported by renders and, in some cases, augmented reality simulations. Proof of concept, from the point of view of the user, requires a 3D scale model or a virtual reality simulation, while "*performance tests*" are satisfied with structural calculations, soil mechanics studies, architectural plans and engineering. For the development of software programs, recognizing user requirements through "*mules*" or "*mock-ups*", we proceed to the construction of horizontal prototypes to materialize the interface of the different screens and to simulate the interaction with the user. Then, proceed to the first stages of development using vertical prototypes to achieve an unstable version or Alpha. When the problems of programming have been Partially solved, there is a Beta version for tests with users or "*Beta testers.*" Other phases of program improvement are often called, continuing with the Greek alphabet, Gamma, Delta, Epsilon and so on until the final version, Omega or RC, is released.

A particular case is the design of services or experiences. The dirty draft or prototype takes the form of a list of characteristics to include in the experience of the client, contemplating the profile of the people who have to attend them, or the technologies for self-service. A representation is built to materialize the experience in the form of a video, a simulation, or a simple render, where spaces, architecture, colors, the use of logos, the dress of the hosts, can all be visualized. Proof of concept requires a mock-up of real dimensions to test the concept with potential users and make the adjustments prior to a formal opening of the service, be it a hospital, a shop, a restaurant or any other possibility in the wide range of products in this category. The integration

Figure 7.
Source: Beristain, L.M. & Moreno, T. (2017) Prototype by the objective-fidelity relationship.

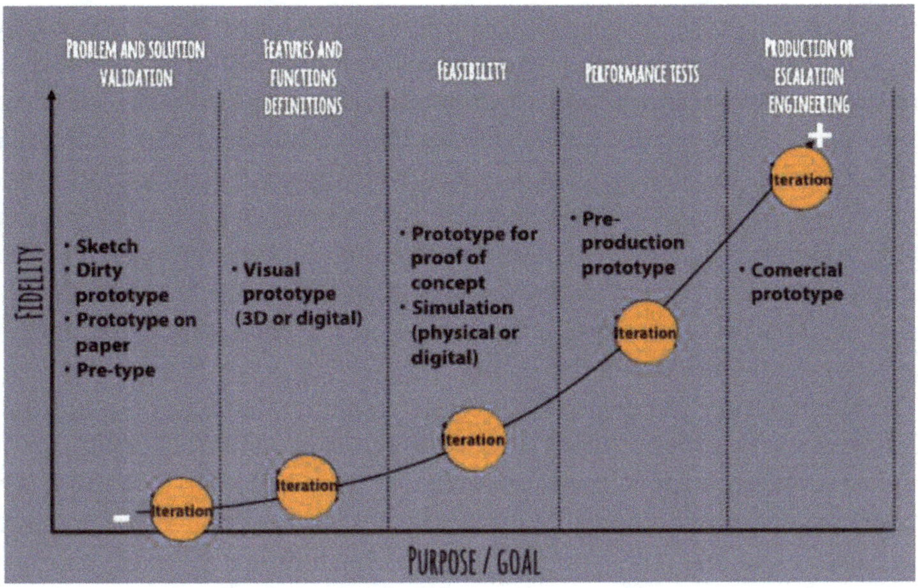

of physical spaces, furniture, technology, processes, and the interaction between the client and the host—employing ethnographic methods—make the service one of the most demanding in terms of investment and commitment developments for a company of any nature. One of the recent iconic cases are the retail Apple Stores, reinventing the way a person buys and lives the experience in a store. The prototype for concept and performance tests took the form of a 1:1 scale model, where it was possible to experience the layout of the products, the displacement of the clients, the architectural design of the spaces, the environment and temperature, and even the packaging of the products, allowing to make the necessary adjustments as a result of the different iterations (Chazin, 2012). The development took two years before the opening of the first store in the Tysons Corner Center of Virginia, becoming one of the most successful chains with about 500 stores in 20 countries (InnvestorApple, 2017). The sale-per-square foot is more than double the luxurious Tiffany's (Chazin, 2012).

Prototype vs. Minimum Viable Product (MVP)

In recent years, design thinking and lean entrepreneurship methodologies have gained popularity by revealing innovative solutions. Multiple programs have been designed from both, aiming to renew the value proposition of established companies or the development of one for the sustenance of the company in gestation. While the first part of a user-centered approach involves addressing the task with a qualitative orientation based on ethnographic observation, the second part focuses on the client with a quantitative approach based on traction and sales metrics. However, the main difference between the two lies in the approach towards the prototype, in the case of the first and the minimum viable product in the second.

Beginning from the need to develop empathy with the user, design-thinking offers the possibility of discovering the innovative concept from the understanding of the person, the context, and the problem to be solved. Any idea or concept must come to life in the physical world through fast and dirty prototypes, including post-it notes; thus, favoring the creative and exploration processes, testing the solution and refining it in an iterative process between the designer and the user. Multiple ideas can be validated without investing too much in time and money.

Lean entrepreneurship, on the other hand, is a method to discover and validate the market for a product, to develop the correct characteristics of the product, and to discover the appropriate channels for acquiring and converting customers. In order to test the different hypotheses, experimentation uses a viable minimum product or *MVP (*Al Ries visualizes the MVP as a tool of experimentation and learning, while Ash Maurya proposes it as the ultimate validation: is the client willing to pay for the solution? Maurya's approach is that of the present document.*)*: The minimum solution that solves the customer's priority problem and produces sales revenue. Strictly speaking, it is not a prototype, but a *real* product; good, simple products to validate the development of a market.

To avoid confusion, a demo, video, or landing page does not meet the MVP criteria adopted here. Ideally, the MVP acquires the form of product concierge, a sample or a pilot production. Other possibilities are the pre-sale and crowdfunding of potential users.

Please use the following examples to clarify the nature of the MVP:

- A little over three decades ago, a gestating company produced 2 kilograms of a coating for aluminum in the kitchen and with the pots of the mother of one of the founders, to test with potential customers. Thirty-two years later, the company employs about three hundred employees and has one of the most modern plants in Mexico.
- In the same way, the first Marinela Gansitos were sold without wrapping, on a waxed red paper, placed in a tray with transparent lid.
- Mobo, the leader in accessories for mobile devices, started in a modest place selling cloth or woven covers for cell phones of the time, at the end of the 90's.
- Metco, the leader in natural sweeteners, started by offering bulk sugar in paper bags in 1990.
- Before building an electric car, Tesla Motors' Straubel transformed a traditional bicycle into an electric one.
- The first 5 personal computers of Steve Jobs and Steve Wozniak, from aesthetics, were crude crafts.

Which is preferable, the prototype or the minimum viable product? Both are validation mechanisms for the existence of the problem and the solution (the product). The prototype requires less resources and time for its development, at least early sketches and visual modeling; The MVP on the other hand, demands more time and resources, but reduced compared to the development of the final product. In both cases, the criterion of failing early, at low cost and learning, to re-iterate is satisfied. Of course, once the iteration with the customers has been introduced to the market and facilitated, the MVP has to evolve to new designs that can not be imagined by using the prototypes. For the established company, it is a matter of selecting the most appropriate method for your business formula renewal project, while for the company in gestation, it is a matter of survival and discovery of a business formula, recommending the use of the MVP.

The light dissemination of the concepts behind design thinking and lean entrepreneurship, as well as their application in group sessions and fields, has favored the diffusion of scientific thought: Any assumption is a hypothesis that must be put to the test. However, the results are poor in assuming that product development is solved in both cases. According to the model proposed here for the design process, both methodologies in their "commercial" form, satisfactorily attend the exploration, discovery, and ideation phases, and partially, the first construction. But from that point, they ignore methodologies for the development of new products and technologies, leaving the entrepreneurial team or the company's collaborators in charge of the innovation project, with no more tools to take the next steps, particularly when access to knowledge of the relevant technology is available. Fortunately, these problems are solved through transactions with public and private agents such as research and development centers, experimental laboratories, universities, technology companies, and open innovation platforms.

Final Considerations

Customers make decisions daily, condemning some products to disappear and favoring the reproduction of others and with them, the companies that originated them. The company has several business formulas (product-market), each one at a different stage in its life cycle. Those with signs of erosion in their sales or performance should be abandoned and replaced by new formulas, if you want to renew the vitality of the company. The key to finding new business formulas is the search for relevant designs through trial and error, characteristic of evolutionary systems, extending the definition of design to everything created with a purpose—as in nature of all the objects created by the human being. For the company in gestation, it is the search for a business formula that allows it to be born and to develop. Along with design and selection, competition is another feature of evolutionary systems. The competition is given in scarce resources and, in the case of the market, the resource refers to the customer's demand.

Discovering and developing relevant design is no easy task. Today, even the simplest product can be integrated by different individual pieces, technologies, and materials. Each piece forms modules together with other pieces, which, in turn are integrated into the system of the product in an architecture. This provides unimaginable exploration opportunities if it is considered that the change can occur in one or more parts, in one or more modules, in the architecture or in the total system. A product with 10 pieces, has 1,024 possible permutations without considering its interaction with the systems of reproduction, distribution, and consumption.

Then, the prototype is the instrument that gives the organization or team of entrepreneurs, with the capacity to experience quickly and at low costs, the assumptions behind an idea or a concept in an iterative process of trial and error in the search for a relevant design.

REFERENCES

Abernathy, W. J., & Utterback, J. M. (1978). Patterns of industrial innovation. *Technology Review*, (June/July): 41–47.

Alcocer, O., & Campos, J. (2014). *The format of convenience stores as a means of proximity of services and supply in urban areas*. Quivera.

Appanniecom. (2017). *Retrospective: Mobile's continued momentum.* Available from: https://www.appannie.com/en/

Cinepoliscorp. (2017). *Cinépolis*. Available from: http://www.cinepolisjobs.com/corporativo/

Designsprint. (2017). *Wwwgvcom.* Available from: http://www.gv.com/sprint/

Designthinking. (2017). *Ideocom.* Available from: https://designthinking.ideo.com/

Femsa, C. (2012). *FEMSA*. Available from: /es/negocios-femsa/femsa-comercio

Gasca, L. (2014). *El libro del fracaso*. Instituto del Fracaso.

Jean-baptiste, S. (2001). *A treatise on political economy; Or the production, distribution, and consumption of wealth*. Batoche Books.

Oxxoinmuebles. (2017). *OXXO*. Available from: http://www.oxxoinmuebles.com.mx/

Porter, M. E. (1998). *On competition*. Boston, MA: Harvard Business Review Books.

Savioa, A (2011). *Pretotype it: Make sure you are building the right it, before you build it right*. Academic Press.

Smith, A. (2000). *An inquiry into the nature and causes of the wealth of nations*. Renascence.

Statistacom. (2017). *Statistacom.* Available from: https://www.statista.com/

Tomke, S. (2003). *Experimentation matters: unlocking the potential of new technologies for innovation*. Boston: Harvard Business School Press.

Ucker, P. F. (1988). *Innovative innovation and entrepreneurship: practice and principles*. Hermes.

KEY TERMS AND DEFINITIONS

Gestating Companies: Term used when referring to startup companies.

Innoduct: Word composed of the terms "innovation" and "pipeline," equivalent to the English language pipeline.

Product: The solution to a client's problem by taking tangible and/or intangible designs into action. If merged, they offer an experience to the client and offer no distinction between product and service.

Sine Qua Non: An indispensable condition, element, or factor.

User/Customer: For purposes of this document, the customer is the person or institution that makes the payment for the product; the one willing to part with a part of its patrimony or credit, to acquire the good or service. It is distinguished from the user and the decision taker. These functions being in a single agent or in three different ones.

Chapter 6
Validation

ABSTRACT

Stragile, like customer development and the methods derived from it, adheres to the principles of the scientific method. When it is said that an idea has been validated, it means that the results are empirical, measurable, repeatable, and falsifiable. Moving from an idea to reality requires moving from concepts to data, from speculation to facts, through observation and experimentation of any innovation proposal or initiative. Organizations and entrepreneurs must have a systematic method to experience new products and concepts on a routine, formal, and disciplined basis. In this chapter, the authors propose that failure in the results of the experimentation is good when it happens quickly and at low cost.

INTRODUCTION

Where do good ideas come from? The vast majority of people want to be more creative, develop better concepts, and belong to innovative teams and organizations. Unfortunately, there is no route or interface to access Platon's *topus uranus*; the place where pure ideas and perfect archetypes exist. In return, the human being has a natural inclination to empathy, allowing him to recognize problems that are worth solving, focusing his attention to produce ideas of arbitrary solution. Some were capricious, others *"clever,"* and some brilliant. But for life and for business, any of them have the same value: none. Only when ideas are set in motion, incubated, and tested, they evolve to the level of an actionable, executable, and validated concept, acquiring value. Between the idea and the execution, there is experimentation.

DOI: 10.4018/978-1-5225-4023-6.ch006

In the previous chapter, the importance of swiftly experimenting with new concepts was discussed; in most cases using the prototype, as an instrument par excellence, to test the assumptions behind a promising business idea. In this chapter, the focus is on actual experimentation, without being able to subtract the influence of the prototype in its execution from the hypotheses of the project.

At the end of 2007, Grupo Elektra announced an agreement with First Automobile Works Group (FAW Group), to commercialize automobiles of the biggest automobile company of China. The target market was new customers for the automotive industry. The agreement contemplated the construction of an assembly plant in Michoacan, México for 2010, with a capacity of 100 thousand units annually (GrupoElektra, 2007), thereby satisfying the regulation to import cars exempting the tariff. Two years later, according to the 2009 Annual Report and in a standing note, Grupo Elektra stated: *"Derived from the global economic contraction and in the automotive sector and due to the delay in the construction of the automotive plant in Michoacan, his registration as a vehicles producer was canceled"* (GrupoElektra, 2009). In other words, the project was canceled. In the fourth quarter 2009 earnings release, the bold initiative was in discontinued operations, with a charge of 276 million pesos, plus a charge for the same concept in 2008, for 78 million pesos. (GrupoElektra, 2009) *What went wrong?* Without a shred of a doubt, the selected year for the introduction of the FAW was a complicated one for such a strategic move: The fall in car sales to the public was the second biggest issue, only behind the crisis of 1995, with one difference: Then, the collapse affected Mexico exclusively. On this occasion, the fall was in the hands of the North American market (Secretary of Economics, 2012) and the global recession. Under these conditions, the investment flows of the plant in Michoacan began to be delayed, bordering on the Ministry of Economy to cancel the permit for the import of vehicles without tariffs. Other reasons, closer to managing the initiative, were related to Elektra's lack of experience in automobile distribution and the inability of the vehicle to meet environmental and safety standards for the United States market, the main objective for output production of the assembler. In fact, the reasons were of a different nature: it is estimated that the sale of FAW vehicles was in the order of 5,000 units in two years (Sanchez, 2014), a ridiculous amount compared to the 100,000 units planned for the Michoacan plant. The product was for "*new buyers*" of cars, which in the context of the Grupo Elektra, means the first purchaser of an automobile throughout the "*family lineage.*"

For a person accustomed to moving by car since childhood, buying a car is an important but routine issue, but for the first-time buyer in the history of a family, it is often incompatible with the client's own value and consumption system and its immediate social environment. However, at the heart of failure is always learning: The customer of Elektra did not want a car, they wanted a motorcycle. Once Grupo Elektra discovered this, it was a matter of focusing on becoming the *"relevant design"* for the market segment and product category. Currently, Italika, the motorcycle division of Elektra introduced in 2004, dominates around 70% of the market, with annual sales in the order of 400,00 units. They have a plant in Estado de Mexico with a production capacity for 500,000 units per year and a network of service centers above 700,000 with a promise of delivery of the required spare, if not in existence, of 78 hours (GrupoSalinas, 2000).

In the study of business formulas, strategic movements, the value renewal and innovation, exogenous causes are always identified (the global recession and the crisis of the automotive industry as in the case described). But also, endogenous causes are identified related to execution, poor preparation, indulgence, but in most cases, by a decision made based on mistaken paradigms and assumptions without question. Despite the crisis, Elektra was able to prop up the motorcycle business, like many other examples of business formulas being born and strengthened precisely in difficult times. However, the Elektra-FAW project proved to be a very costly experiment, in the same way as 75% of the projects that cannot survive the two years of operation. In business, there are no triumphs or failures, only experiences... some expensive and others quite reasonable.

As quoted in the previous chapter, Adam Smith conceptualized failed experiments as a cost to the performer, but successful ones as a profit for the whole society. Likewise, for experiments are run with greater or lesser risk, and not always with a return to the experimenter. But all economists agree: Without experimentation, there is no innovation and, consequently, no economic growth. This being so, it is advisable to design quick, simple, and inexpensive experiments, as far as possible.

Scientific Thinking Applied to Business

After analyzing hundreds of projects comparing successful to failures, it is possible to recognize patterns such as the effective use—or absence—of the scientific method applied to the development of new products or to the creation of new companies. This is characterized by assuming an explanatory condition of the phenomenon under study. From the relationship established between variables, the consequence or effect between them is deduced, verifying by means of experimentation or observation, the facts that validate or not validate the prediction (Gower, 1997). In sum, the scientific method is knowledge based on empirical and measurable evidence.

As discussed in the previous chapter, managing evidence-based innovation is not a new approach. The creation of utility of resources requires to be experienced and tested as asserted by Adam Smith in 1776 and Jean-Batiste Say in 1880. Edward Deming proposed the PDSA (plan-do-study-act) circle in 1950 for development of new products with the integration of teams from the areas of design, production, and marketing with the aim of quality and recommending small-scale experimentation, as proposed by Drucker in his work on innovation and innovative entrepreneurship. Stephan Thomke published in 2003, the results of his research on new experimental techniques, enabled by emerging technologies such as computer simulation and 3D printing. Recently, the use of the scientific method has been popularized thanks to the 'Lean Startup' movement, promoted by Eric Ries on 2012.

Thus, innovation or problem solving generally starts from an idea; from a concept—which traditional literature calls the "identification of opportunity"—whose form is that of a hypothesis of the type: If A, then B. In general, the authors of the lean enterpreneurship recognize at least two fundamental hypotheses: (1) The hypothesis about the *problem*, also called the penalty, the need or the desire of a segment of customers or the population, and (2) the *solution* or product hypothesis, to solve said problem. In practice, each of these original hypotheses is composed of several sub-hypotheses and many more are required to reach the validity condition. Generally, in the field of business, no assumptions are made to support a decision. The problem with the assumptions behind an idea, a business plan, or a proposal of any nature, is the construction based on motives given by a degree of certainty or truth without any demonstration. Instead, the hypothesis is established for the sole purpose of being falsifiable or subject to refutation by new data. And if something constantly changes the data, it is the environment of the company.

The scientific method applied to business is nothing more than the requirement to move from assumptions to facts in the set of motives behind any business opportunity, innovation, or entrepreneurship where observation and experimentation are the tools for discovery and learning of the market segment, the value proposition, technology, channels, mechanisms of income generation, price or the structure of costs, and in general, of each of the elements of the formula of business. Discovery and learning must always be validated by observation and translated into hard data, converting assumptions into facts to provide a foundation for the proposal.

The Procedure of Scientific Thinking Applied to Business

When a piece of data (observation) does not fit the assumptions about a market, a product, or a process, the company man or woman investigates thoroughly to understand the anomaly. *Why have sales stalled? Why did such a customer stopped buying from me? Why is the best selling product being replaced by another? Are there any completely different value propositions in the market? What do my competitors do? Why can such a competitor offer better prices to their own?* Endless questions that should cause the alert man or woman, noting that it is not in line with the assumptions of the company and possibly the industry. This involves going out, observing and asking, to customers, not-customers, competitors, suppliers, collaborators of all levels, specialists, and even external to the cluster itself, trying to understand if the original assumptions have lost their validity.

In this first stage, it is useful to know the censuses, studies, and samples on a topic, to validate some of the assumptions. As a general rule, it is not necessary to collect data if they have already been collected, as long as the source is recognized, the methodology satisfies the research criteria, and the data is relevant to the search initiated. If the data does not exist, a survey is applied to a sample (quantitative) or the behavior is observed by ethnographic methods (qualitative) or indirectly derived from other existing data. This procedure corresponds to the observation phase to answer the question that initiates the process. However, no matter how much evidence can be gathered through observation, it will always be necessary to test the idea through operations aimed at discovering, demonstrating, or testing market and technology assumptions, irrespective of the magnitude of the investment required for its execution.

From the gathered information, the entrepreneur or businessperson begins to construct a hypothesis of the type, *"If A is true, then B will be true."* The interesting thing about the approach is that there is a possibility that A is false and B is true. For example, let's assume that "*if A is true*" is an investment in a manufacturing team and "*B is true*" is the profitability of the project. It is understood that if A is true, it will not necessarily be B; and even in the process could find another formula that makes "A" false, but "B" true.

The wonderful thing about the hypothesis is its structural disposition to be falsified or denied with, and only with evidence.

Once the hypothesis has been established, a second stage is designed to test it, clarifying the learning objectives, the assumptions to be validated and the best instrument to run the experiment. The main difference between observation and experiment is the *recognition of associations* in the first and *causes* in the second, assuming that it is possible to control the variables.

The vast majority of experiments in business require the construction of a prototype, to be tested under controlled conditions such as surroundings, context (real or simulated), environment, and methods for data collection.

Resolved the above, proceed to run the experiment, collect the data, and analyze the results in detail. The conclusions can validate or not the assumptions behind the hypothesis. In the first case, the decision must be taken to continue with the following steps of the project. In the second, to provoke a new hypothesis; to repeat the process. Figure 1 summarizes the scientific method applied to business, innovation and entrepreneurship.

As in inferential statistics (establishing the relationship between variables for a research or work hypothesis), the same process is used for the agile strategy for it creates connections for each of the components of the business formula.

Figure 1.
Source: Beristain, L.M. & Moreno, T. (2017) The scientific method applied to business.

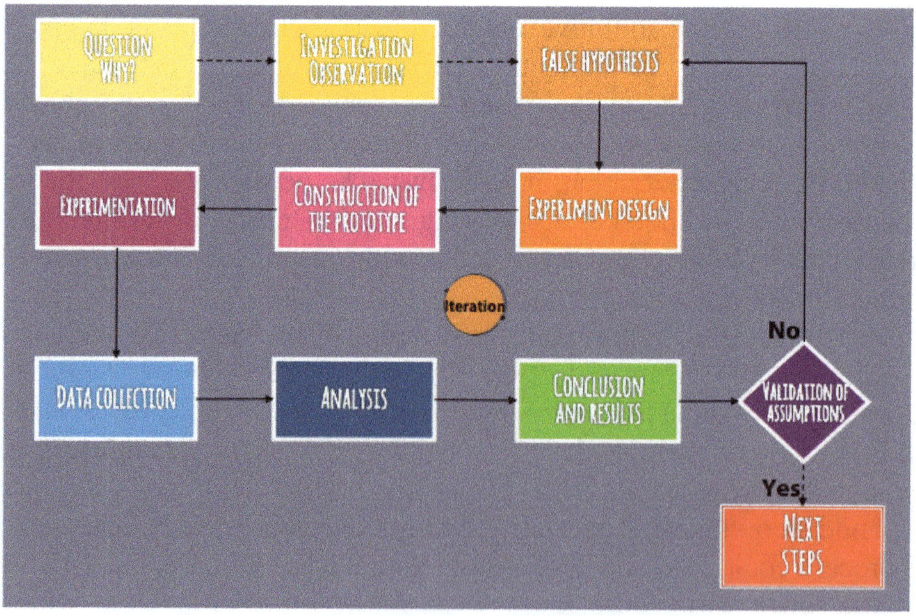

When the results do not reject the null hypothesis (the one where there is no relation between variables, equivalent to not validating the assumptions), the researcher returns to his original approach to reformulate its substantive nature or abandon it on the basis of being impossible or absurd. On the contrary, to reject the null hypothesis advances to the following stages iteratively in the process of shaping the body of knowledge achieved on the basis of evidence. Similarly, the innovator and the entrepreneur must approach the different hypotheses of the project through experiments and tests of different nature. Typically, it is necessary to iterate until validation is achieved before continuing with the next phase of the project and sometimes it is necessary to make substantial changes or to pivot:

A pivot is a substantive change in one or more of the nine boxes on the canvas of the business model...The iterations, on the other hand, are minor changes to the components of the business model. (Blank & Dorf, 2012).

...particularely, when the focus is on the client's central hypothesis of problem and solution. (Cooper & Vlaskovits, 2010)

Unlike inferential statistics, the procedure does not follow a rigorous method of measurement, but rather a rigorous method of learning based on feedback from potential clients. Also, the objective is not to reject the hypothesis of work, but to iterate or to pivot until discovering the pertinent relation for the formula or the business model. Put another way, the goal is to discover a business formula based on empirical and measurable results through iterating and pivoting.

Stragile refers, essentially, to the discovery of a business formula validated empirically and in no other way, and its subsequent scaling through a process of verification of the hypotheses and assumptions of the project. Table 1 summarizes the set of basic hypotheses—the elements that, until they are not validated, continue to be assumptions or conjectures—for each of the stages: In the early stages of development of the company in gestation or of innovation in the established company, according to Ash Maurya (Mauraya, 2012).

Three warnings about the hypothesis: First, it must be designed to be falsifiable, verifiable and repeatable. Consider the following wording for the first of the image (there are Z people with problem X, willing to pay for a solution): "*4 out of 10 professionals under 40 years of age in urban areas devote between 46,800 and 78,000 Annual hours in your car to move to or*

Table 1.

	Step 1: Problem-solution fit	Step 2: Problem-Market fit	Stage 3: Scaling
Hypothesis	There are Z people with problem X, willing to pay for a solution.	There is a MVP of R producing sales.	The formula / business model is repeatable.
	The best way to solve problem X, is with R.	The best traction channels are A, B and C.	
		The growth engine is (viral / paid / retention).	The formula / business model is scalable.
	People with problem X consider the R solution as a "must have".	The sales price less the cost of acquisition of customers, less the cost of the product, results in a gross margin of Y%.	
	People with problem X are willing to pay $ x for solution R.	There are reliable suppliers to meet a demand for Q units daily.	The formula / business model is automated.
	The technology of solution R is accessible and ready (vs development).	MVP technology works within established standards.	The formula / business model is protectable / defensible.

Source: Bersitain, L.M. (2017).

from work to home "(known as" workilometers"). With some creativity and effort, it is possible to deny the claim with verifiable and repeatable data—a survey properly designed to be applied in professional social networks such as Linkedin, should resolve it. Another possibility is to ask anyone who meets the profile in a transit zone, such as a shopping plaza or an avenue, and record the answers. Once the first assumption has been validated, the next step is to design the experiment for the second assumption in the first column. (The best way to solve problem X is with R), requiring the construction of a dirty or pre-prototype of the idea of solution, and leave new account to ask always to people with the profile of the customer (in fact, at this point, even the customer profile is a hypothesis). The hypothesis can be written as follows: "*Professionals who spend between 46,800 and 78,000 hours per year to go to or return from work are willing to rent electric scooters to travel the last mile, where they lose 30 to 40 minutes looking for parking or in the congestion caused by the units looking for parking." Again, the hypothesis is falsifiable, verifiable, and repeatable. The writing of the hypothesis requires a market segment (professional "workilometers")*, an observable and verifiable action (renting an electric scooter), and a reason for decision or reason (waste of time searching for parking).

Second, it is necessary to prioritize the risk-based hypotheses, always starting by validating the existence of a market willing to buy the solution at a certain price, continuing with the validation of the technological or administrative feasibility, and finally, validating the execution capacity. Not long ago, a family had begun a large investment to rebuild a property and put it up for clinical care services: the purchase of the building, the architectural designs, the market studies, but they had not validated if the corresponding authority would grant the permits for functioning. In this case, after validating the market, it is convenient to solve everything related to the "*technology of the solution*", so try an administrative authorization. The risk of the project is "*moved*" from one place to another as progress is made in the development of the project. The riskiest hypothesis is always the first to test (usually), the existence of a market.

Finally, the hypothesis reference of the image is not, even remotely, exhaustive. Each project, with its respective endowment of resources (particularly of talent), must identify which elements are assumptions and which are validated data; what is an illusion and what are facts. Do not be fooled by the obvious or the method relaxation. Even Steve Blank, the initiator of the whole movement related to lean entrepreneurship, conjectured: "*There are businesses where the aphorism, 'if you build it, they will come' is indeed true. For example, in biotech entrepreneurship, if you can find a drug to cure a specific cancer, a Customer Development strategy is simply a waste of time.*" (Blank, 2007). The assertion can actually be interpreted as an acknowledgment of the relevance of the product development process when the technological risk is greater than market risk. The assumption can not be taken for granted when hypotheses to be confirmed are missing. Example: If the drug cures the specific cancer, (1) *would customers buy it*? (2) *Would they buy it at that price*? (3) *Would they buy it from our company*? Or, (4) *are the side effects relevant*? And many more questions.

The Design of Experiments

Designing an experiment is nothing more than planning the procedure to follow to produce the correct data in sufficient quantity, with the intention of answering the research questions. These, from experience, are recommended to be grouped into those related to (1) the problem and the client; (2) solution and product development; (3) with traction channels and growth engines, and (4) with the other elements of the business model and the development of the organization, particularly strategic activities and resources to replicate the relevant design.

In the experiments of the natural sciences, the investigator manipulates the independent variable (cause) to observe the change of behavior in the dependent variable (effect or result). When he can not manipulate them, he proceeds with observation and exploration. Regarding the development of the solution (the product), the means of experimentation in science and engineering can be largely invoked. However, in the case of socio-economic systems such as the market, the client, the environment, and the supply, the experimentation space is extremely complex: Sometimes the variables themselves are unknown or, at best, uncertain; in other cases, they change with some frequency (consider the exchange rate of a currency or interest rates) and the magnitude of the relationship is rarely known as the causality between them. That is why we must proceed with caution, in a tentative way, and in trial and error mode.

In order to prepare the experiment or the observation, the learning objectives must be established and the working hypothesis formulated. In fact, the definition of the hypothesis largely determines the nature of the experiment. Consider the previous example: *"4 out of 10 professionals under 40 years of age in urban areas devote 46,800 to 78,000 hours per year in commuting to and from work."* In the first stage, investigate whether there are already measurements or data that validate the assumption. If they are not available or are not convenient, then the best way to do this is by means of a survey, exclusively for this purpose (not asking for anything else), applied electronically or in person, of at least 383 random observations to satisfy the criteria statistics (5% error, 95% confidence and population > 100,000). Unfortunately, the observation is not without error, particularly in the design of the survey or in the conditions of application. If the interviewer makes observations in a central office area, most respondents are likely to spend that time moving between home and work. However, if performed on a holiday or on a Friday, with rainy or dry weather and multiple uncontrollable variables, the results will vary. Within possibility, take into account any forecast to avoid Type I errors (rejection of the null hypothesis being true in the population) or Type II (failure to reject the null hypothesis being false in the population).

Validated the previous assumption, in a second stage we proceed with an experiment based on the hypothesis: "*Professionals who devote between 46, 800 and 78,000 hours per year to go to or return from work, are willing to rent electric scooters to move the last mile, where they lose 30 to 40 minutes looking for parking or in the congestion caused by the units looking for parking*." By their nature, hypotheses whose dependent variable consists of an intention or decision, should not be addressed through surveys; at least not in business—and recent experience indicates that neither in politics. Over the years, multiple cases and examples accumulate, documenting efforts with the intention of resolving such assumptions through surveys, particularly when it comes to innovative products. Questions like, *would you be willing to park your car 8 km away from your destination*? Or, *are you interested in a recorder that does not record?* invites a negative response without further reflection. In 1978 (Akio, Reingold & Shimomura, 1988), Masaro Ibuka and Akio Morita, co-founders of Sony, decided to develop the first portable cassette player, modifying the company's portable tape recorder. To achieve the dimensions and weight suggested by the founders, it was necessary to remove the system to record. In the opinion of the engineering and marketing staff of the company, even the retailers themselves, the product had little chance of success, who would want a recorder that does not record? In fact, market surveys showed this. But Morita continued with the project, building the first prototype in 1978. The market tests with this one were a success and for 1979, they launch the Walkman, being the second most successful product in the category, with 385 million units sold until 2009 (Date it was discontinued) (Sony, 2009), very close to the 390 million iPods reported to 2014, the latest figure published by Apple (Applecom,2015). As Morita himself said:

I do not think any number of market surveys would have let us know that the Sony Walkman would be a [failure or a] success, or say a sensational hit that would generate many imitators. (Morita, Reingold & Shimomura, 1988).

Let us continue with the electric scooter rental example. Before making any investment, it is appropriate to construct a dirty or pre-type prototype, such as a drawing, presentation, plastic model, digital mockup, video, construction with various Materials and colors— pieces of plastic, wood, cardboard and even plasticine, to mention some possibilities—to validate with the client or potential user, the existence of the problem and the functionality; characteristics and elements of the solution. Through several iterations, the innovative concept is refined. At this point, even in the case of digital products, it is advisable to proceed face-to-face in order to capture not only words, but nonverbal reactions and behaviors. The enthusiasm reflected in the face, the strangeness or the disapproval are all valuable feedback. A focal group is also not recommended, to avoid cross-contamination in feedback. The most frequent question of the teams responsible for an initiative of innovation in the company and of the entrepreneurs is, *what is the adequate number of people to whom to present the concept and to receive feedback*? From experience, after 6 or 7 in-depth interviews, 80% of the valuable information for the project will have been discovered, stopping when each additional interview does not bring new data or information, following the principle of decreasing marginal returns of the experimentation.

As proposed by design experts, the dirty prototype should be constructed as if the concept was correct, but the validation test run the as if one were wrong. In the worst-case scenario, the initial concept may be wrong, but with the possibility of evolving to the right solution—if the problem is to find parking near the office, possibly the solution is not to rent scooters but to develop an "*Airbnb*" from the parking drawers, or find out if Uber has solved the problem, or if a bicycle is a preferred choice over the scooter, and many other possible satisfactions with their respective combinations including the non-existence of the problem. The objective of this experimentation is to refine the understanding of the problem, to better understand the potential customer and the user, and to know if the solution approach is correct, debugging its characteristics and exploring the price that the customer would be willing to pay if it were the case.

Without going into details about the statistical test of hypotheses, to validate means to confirm the degree of probability of the data or the relation established in the hypothesis, or, to get to know it. In the previous example, about 40%of the cases, the '*workilometers*' allocate between 46, 800 and 78,000 hours a year in commute from home to work and back, or, to discover that this percentage allocates 5,000 hours more or that the percentage of 'workilometers' is lower or higher. In the same way, it must be validated if the time spent, from 30 to 40 minutes) are spent exclusively on the 'last mile' seeking parking or moving in the congestion caused by this reason. Likewise, the proportion of '*workilometers*' arranged to rent an electric scooter for the last mile, as a function of the time spent daily, is confirmed. Possibly a person who devotes more than 3 hours a day is indifferent about saving 30 minutes, being a more attractive notion for those who spend less than 3 hours of transportation every day.

If the results obtained justify it, proceed in a third stage with the following set of hypotheses. To confirm the scooter's rent intention and the price willing to pay for potential customers, something more than a pre-type is required. At this point, a verbal manifestation of intention is not enough, but a compromise of choice between the multiple offers, recommending the use of a minimum viable product (MVP). For the example developed, this means buying or renting a pair of scooters, processing of traffic permits, acquiring of accident insurance and civil liability, temporary negotiation of a pair of parking spaces in a lot on the appropriate route, enabling the energy sources for the load, the platform of customer service and the different arrangements necessary to be able to perform the test, offering the service in real conditions. Of course, depending on the nature of the product and its technology, building a MVP may require (1) a significant investment to cover development costs, (2) attention to regulations generally designed for customer safety, and (3) a careful approach if the product influences consumer health. Here are some examples to clarify the concept of MVP:

- For the development of the Model S, Tesla Motors built the prototype by modifying and converting a Mercedes CLS into a fully electric car (Vance, 2015).
- Before building an assembler for the Italika bikes, Grupo Elektra bought the design and a pilot production from the Korean company Hyosung (Zorrila, 2015).
- In the pharmaceutical industry, the prototype drug is required prior to pre-clinical and clinical studies, meeting all the requirements of the current regulation for experimentation in living beings (PHRMA, 2016).
- The architectural project of the new International Airport of Mexico City (Aeropuertogobmx, 2015), integrated a MVP with drawings, plans, perspectives, and engineering, accompanied by renders and physical and digital models.
- Henry Ford's first vehicle was the "quad," handcrafted and sold for $200 USD in 1896 ($ 8,668 in 2016) (TheHenryFord,2010).

In other cases, it means a production of two kilos of coating for aluminum, the preparation of cakes filled with cream and jam without wrappin, woven fabric covers for mobile devices, or bulk sugar in paper sachets. Of course, we also consider the apps or digital platforms of the concierge type or an irresistible function. In most examples, except for pharmaceuticals, the two basic criteria of the MVP are met: (1) Solves the customer problem and (2) produces sales revenue. In the case of pharmaceuticals, it surely solves a problem and, being successful in the previous tests, will attract sufficient investment to finance the subsequent studies.

A good experimental design with the MVP should provide data and information on various hypotheses related to the problem, the client, and the solution. For the latter, through the feedback of potential customers and various iterations, the characteristics, functions, and expression of the product are refined allowing it to reach a degree of development sufficient to become a "*must have*" solution and motivates the client to pay for it. At this point, we proceed to confirm or discover other assumptions of the business model, particularly the price that the customer would be willing to pay for the solution, traction channels and engines of growth. One begins to confirm the fit between the product and the market when prospects are acquired, they experience the product, repeat the experience, and pay for it. The customer conversion rate and its retention will determine if you are ready to continue with the stages of organizational development and escalation, or if it is necessary to continue the iteration and pivot process until you achieve the product-market adjustment.

The products integrated by a set of services such as the case of a restaurant or an ambulatory care clinic require a special mention. In these, the concept must be validated in its entirety and for each of its components. For a coffee shop, the value proposition must comprise the quality of the product, the atmosphere, and the service in a concrete and integrated way; as well as for each grain variant, for each form of preparation, or for each size. An outpatient surgical services clinic requires validating each specialty and with it, the necessary infrastructure. But the value proposition should additionally include the profile of the human resource, the quality and safety of the services, the certifications and regulations, and of course the location. Although the Stragile process is the same, the definition of the MVP is of greater importance and, at the same time, it results from greater complexity and investment.

It is not enough to say that, in the case of a multi-hypothesis experiment, it is convenient to perform the necessary ones, manipulating one variable at a time while keeping the rest constant. As we proceed with the validation of each of the assumptions, the related variables are "frozen" and we proceed to the iteration of the remaining variables. The final objective is to validate the hypothesis set of stages 1 and 2 described in the image.

At this point, it is convenient to remember and warn that Stragile is a way of thinking that integrates various methods to innovate in the business formulas of the established company or in the search for a formula of business that of life to a company in gestation. By definition, the new business formulas result in innovations of a disruptive or discontinuous nature. On the other hand, when the innovation proposal is a result of planning in a technological roadmap (technological pipeline) or is adjective and incremental, it is recommended to use the best practices and methods for the development of new products the industry. Disruptive changes occur in contexts of high uncertainty and risk, while incremental changes do so in predictable and stable environments. The Stragile method allows, through experimentation, to produce enough data and information to calculate the probability of success of an initiative and, in some cases, transfer the experimentation to the scaling using the action-research approach, thus reducing the degree of initial uncertainty: With data, it is feasible to estimate the odds of success of the initiative. Figure 2 summarizes the validation process proposed in this chapter.

Repertory of Experiments for Consumer Products (B2C)

In order to validate the different assumptions of an innovative initiative, there are different experiments with the ability to test the set of hypotheses, particularly if the client is willing to pay to solve the problem, how much he is willing to pay, if the solution is preferable over other solutions, and if it has the minimum characteristics to create value. In these cases, the prototype usually takes the form of a minimum viable product, one that solves the customer's core problem and produces revenue from sales.

Figure 2.
Source: Beristain, L.M. & Moreno, T. (2017) Process of validation of the innovative concept.

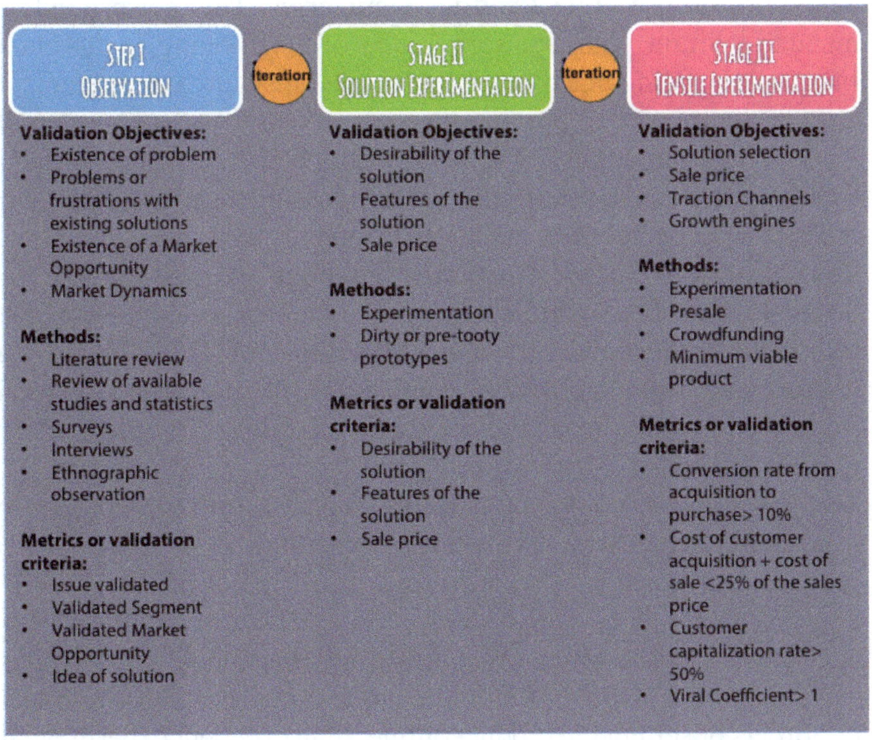

- **Presale:** Frequent practice in some industries such as construction, but equally applied when the expectation about the product drives the demand of innovative consumers, such as Tesla Motors Model 3 (TheTeslaMotorsTeam, 2016) or Glowing Plant fluorescent plant seeds (Glowingplantcom, 2016).
- **Sponsorship Crowdfunding Campaign:** Modality of capital formation requested from potential customers in the form of seed capital collected through online platforms. It allows validating the public interest in the solution, business concept, or activity that is proposed, favoring the approach to future clients willing to sponsor the development of the solution. Although there are multiple cases, probably the most successful has been the Pebble smart watch, bringing together over $20 million USD from 78,471 sponsors (PebbleTime, 2017).

Sample or pilot batch of the minimum viable product. Several examples have been provided of this modality throughout Chapter 5 and 6 such as the first Apple computers (Shea, 2013), MOBO fabric sleeves (Arce, 2013), Metco paper sachets (Reyna, 2007), or Henry Ford'squad (TheHenryFord, 2010).

- **The Minimum Viable Product Concierge:** This corresponds to a manual solution with the appearance of functioning in its entirety. Imagine a vending machine of soft drinks, inside which is a person to receive the money, to provide the change and to deliver the product, without the user perceiving it. The first page of Amazon consisted of a capture screen and a list of titles ordered by subject or by author, whose orders had to be attended manually (Stone, 2013).
- **The Minimum Viable Illusionist Product *(Wizard of Oz)*:** In its origins, Zappos, the world's largest shoe retailer, tested the concept by selling the product under the name ShoeSite.com. To avoid investment in space and inventory, the co-founders directly purchased the product from a competitor from the locality, satisfying the orders received (Staley, 2014).
- **Test Ads:** It does not matter if the business formula consists of a digital product distributed by digital channels or if it is a physical product distributed in a "*brick and mortar*" point of sale, surely the target segment lives and resides online. The use of AdWords—or alternatives such as Bing Ads, AdRoll or Facebook Ads—to design campaigns that lead to a landing page and invite action by collecting pre-orders or selling the product online, is an excellent experimentation

to validate the effectiveness of the ads, the attractiveness of the offer, the willingness of the customer to pay for the solution, the price, and the effective traction channels.

These six experiments allow us to confirm the starting hypotheses related to the problem, the solution, and the market segment. At the same time, it allows to validate or discover the variables related to the sale price, traction channels, and engines of growth. However, it does not necessarily solve the assumption of the system and cost of reproduction of the product.

How to Proceed When Developing Products for Industrial (B2B) Markets?

When the renewal of the business formula refers to industrial products, it is necessary to build close and trustful relationships between the people who make up the development team of the companies or organizations involved. The market risk is greater than that of consumer products because of the small number of buyers and because of the uses and customs of the industry in the customer-supplier relations. If in some context the aphorism of the *"client as co-creator"* of innovation applies, it is certainly about this.

A little more than a decade ago, high corn fructose threatened to replace sugar in carbonated beverages, putting serious strains on the already battered sugar mill sector in Mexico. According to some studies, high fructose would lower obesity—although it does not contain the same nutritional value as sugar—and, at least for bottlers, the input would cut sweetener costs by 30-40%. *Does not seem like much?* In 1970, 98% of the soft drinks produced in the United States were sweetened with sugar, the rest with fructose. By 1980 the ratio changed to 71% made with sugar and 29% with fructose. But in 1996, 97% of soft drinks used high fructose and only 3% used sugar. According to some experts, given the new technologies and new knowledge of business strategy, what in the United States took thirty years, in Mexico can take only ten (Valdez, 1997). Since NAFTA was in force in 1993, high fructose began a process of sugar substitution, reaching in 2010 a infiltration of close to 30% of the sweetener market in Mexico (Garcia, 2011).

Contrary to what some national publications (linked or promoted by people related to the sugar industry), HFCS (no high fructose corn syrup) produces no harm and is classified by the Food and Drug Administration (FDA) as, *"HFCS is safe for use in food as well as sucrose, corn sugar, corn syrup and invert sugar.*" However, a distinction should be made between crystalline and high-fructose (HFCS). Since the first contains only fructose, while the second contains practically the same amounts of glucose as fructose, similar for practical purposes to sucrose (CalorieControlCouncil, 2006).

Despite all the arguments in favor of HFCS, prices between sugar and high fructose have changed in favor of the former. Although in 1994 the price provided for a sugar substitution in the bottling industry, from 2000 the $160 USD per ton gap between HFCS 55 and refined sugar was reduced until the ratio was reversed towards the end of 2007. Thereafter, becoming the first more expensive for about $40 a ton. In the case of standard sugar, the price is favorable to it for just over $125 USD. Record prices in 2007 are mainly due to the demand generated by ethanol production and the high costs of corn (USDA, 2007), a situation that has remained to date (MacroTrends, 2016). In 2005, ethanol production in the United States was 3.7 billion gallons, reaching 6.2 billion in 2007 and, according to the plans of expansion of the productive capacity, reached the 14.28 billion in 2013. While this is "*good*" news for Mexico's sugar industry, the effect on the price of food derived from maize, which could be aggravated by an increase in the cost of fuel, should be considered (Baker & Zahniser, 2007).

They serve as background to the renewal of a depleted business formula, regardless of factors exogenous to the industry and the policy efforts made to protect the sector. Some entrepreneurs and researchers are convinced that in Mexico you can innovate to be more competitive. This is the case of Federico López Medel de Procazúcar, S.A. Of C.V. And the brothers Perdomo Bueno del Ingenio of San José de Abajo, Veracruz, who in 2002 began work to be able to produce refined liquid sugar and to be able to compete with the HFCS. Just as an antecedent, the bottling industry consumes approximately 2.1 million tons of the sweetener, which must be processed to convert it into liquid, reason why the input is emptied, diluted and filtered, with a reduction not less than 0.5% of the total. The design of the liquid sugar mill made by Mexican engineers for 96,000 tons per year gave surprising results: The investment in the plant was $1.5 million, 20% that of a similar plant in the United States; The conversion efficiency of solid bases was 99%. For the second year of operation, the gain on the project was $4.2 million, while the recovery of the investment occurred practically before the first year. Given the conversion efficiency, the residuals were minimal, being used as input for pig farms or as an irrigation resource in the field, which makes it a non-polluting process. Liquid sugar competes with HFCS and refined sugar at a price similar to that of the latter, but at a cost of production of about $15 a ton, or 30% of the cost of refined sugar. In addition to the simplification of the plant and equipment process design, which is highly automated to the extent that it requires 3 operators and little installation space—this is a "*mini-plant*"—it was innovated in the process of Phosphothlotation, which in traditional plants only discolors up to 30% (Emio national in science and technology in food, 2004) of the sugar fed before moving to the crystallization and spinning stages, while the new technology decolores up to 95%. The new technology is undoubtedly an attractive opportunity for the modernization of the mills in the country and for the bottling and food plants that demand refined sugar in liquid form.

Although the solution is evident, the problem was not clear in its beginning. A whole industry tried to influence the authorities to create tariff barriers to the importation of high fructose, arguing for reasons of health and nutrition—as if the consumer of carbonated beverages was worried about its well-being—as of price (dumping). In practice, no one approached the customer to understand why he preferred high fructose over sugar, as did the owners of Ingenio San José de Abajo and the engineers of Procazúcar: The bottler had to transform the sugar from the mills into liquid for incorporate it into its process, adding unnecessary costs. With high liquid corn fructose, this process was eliminated and costs were reduced. What is pressing is knowing that, once the solution is known, no other sugar mill has adopted the technology.

As discussed in Chapter 5, the exploration and discovery of real customer problems, whether articulated, disjointed, dissatisfied, or unknown; and frustrations with product performance, are the first step in developing innovative solutions. Thereby renewing exhausted business formulas. The innovation of industrial products or business-to-business (B2B), parts from gathering information from the customers and users. Project teams must gather it first-hand, identifying leading users and product experts to understand what the task is to be accomplished and what frustrations to take in the process. Defining the problem with clarity and precision is the first step in order to proceed evolutionarily from the preliminary concepts to the final solution, at least for a while. The idea of liquid sugar was not the first concept, but the result of a process of ideation.

VALIDATION AND FINAL CONSIDERATIONS

Stragile proposes a frame of reference for the experimentation and validation of hypotheses in three moments: a first instance of observation, exploration, and understanding of the problem and of the solution; confirming the existence of the first and the magnitude of the opportunity. In a second moment and armed with a dirty prototype or pretotype, proceed to refine the solution and the minimum characteristics to solve the problem of the client. Finally, experimentation with a minimum viable product in any of the ways proposed here, to generate traction. The movement from one stage to another is not linear, as suggested in Figure 3, but an evolutionary process through iterations, making the decision to continue or abandon depending on the results obtained in each of the experiments.

To validate in this context means to ensure that the relationships established between the variables of the hypothesis occur with a known coefficient or in a range of probability that justifies the market opportunity. It differs from verification that it consists of confirming compliance with norms, standards, regulations, or conditions imposed by the customer, authority, technology or usage requirements. Experimentation allows to gather data and information as evidence of validation.

As a conclusion, it is recommended to observe the following principles when the objective is to discover a business formula based on a relevant design that is repeatable, scalable, and automated:

Principle 1: *Test decisive assumptions and review them periodically.* Since business management and entrepreneurship are practical knowledge, decisions must be a product of evidence and facts, not of business intuition or perceptions. The general understanding (duly confirmed) about the operation of the business, reflects the theory of the company. Each one of the strategic assumptions must be adjusted to reality to be effective. When an anomaly arises, it is necessary to proceed with the exploratory work for the understanding of irregular data, with the subsequent experiments, to either incorporate or not the new concept in the logic of business theory. For Stragile, the recommended attitude to keep the relevance of the company in force is the one suggested by Tetlock and Gardner: *"Test, Fail, Analyze, Adjust and Try Again."* (Lock, Philip & Gardner, 2015)

Principle 2: *You can experience anything.* This statement may sound pretentious and tasteless, but with creativity and using the multiple resources of prototyping, simulation and product development, it is possible—always observing ethical principles and international standards for experimentation with living beings. Thanks to the advancements of technology, the gap between what can be imagined and what can be done has been reduced. *Can a person with spinal cord injury regain movement?* A few years ago, Dr. Grégoire Courtine, a postdoctoral student at the University of California, Los Angeles, proposed to his tutor, Dr. Reggie Edgerton, a novel idea to revive neural networks and restore movement. After listening carefully to the concept, Dr. Edgerton said, "*Why don't' you try it?*" After ten years of exploration and experimentation, the "*neural bypass*" in animals is a reality. With the exception of science-based products, experimenting with products and concepts should be a task that is encouraged by organizations on a recurring and even routine basis, but formal and disciplined.

Principle 3*: Failure is good when it happens quickly and at low cost.* Failing years later and after a significant investment, is not recommended, even assuming slack in resources. In any case, an open, objective, and flexible approach to the results of experimentation is required: Any validation process is subject to multiple errors and erroneous conclusions, so they should be declared in the form of a hypothesis susceptible of being falsifiable; giving the facts primacy over any judgment or evaluation until proven otherwise. Of course, the failure of an experiment should not be confused with the failure of an execution by omission or negligence. In the first, it is a goal, in the second it is a "*fault.*" Likewise, as a rule, the number of experiments in stages II and III—and thus the number of failures—is inversely proportional to the quality and depth of observation work from stage I. "*Doing the task well*," is the basis of a "*good failure.*" Be creative to find ways to spend little on experimentation, but to gain disproportionately in learning.

Principle 4: *Do not confuse learning objective with business results.* The purpose of experimentation is not to refuse or reject a hypothesis, but also to generate an economic result or gain a position in the market. The only purpose is to produce a validated learning. In the design of the experiment, the hypotheses are established as a manifestation of the expected results. Once the experiment is done, the actual results will be obtained. You have to compare: *What happened*? *Why*? The analysis of the differences, of the concordances, and the reasons of both, produce a validated learning.

Principle 5: *Keep everything properly documented.* The documentation should take the form of a log where the team members record the objectives, design of the experiment, development of the hypotheses, execution of the experiment, results, analisis, conclusions, and decisions about the assumptions. Even in the case of small teams for the development of the project, as it is often the case in start-up companies, there are problems of communication and understanding, recommend simple and unpretentious means, but carefully attended to obtain some benefit from learning and make knowledge productive.

Principle 6: *Understand the difference between "discovering" the gold mine and "exploiting" the mine.* We have stated throughout Chapters 5 and 6 that the goal is to discover a repeatable, scalable, and automated business formula. Once identified, the organization has the responsibility to bring it to the next levels of productivity and efficiency. Organizations of this type must be predictable, routine, and bureaucratic. There is no such thing as the "*innovative*" organization or the "f*lexible*" organization, not in the strict sense of the term. In any case, there are teams to explore, experiment, and learn. If they find a product or a promising concept, the work is done to introduce it to the processes and routines of the organization. Each requires different metrics and treatments, resulting in a complex collaboration between the two systems, except when the escalation stage is reached. But the latter is an assumption and must be tested.

REFERENCES

Aeropuertogobmx. (2015). *Aeropuertogobmx.* Available from: http://www.aeropuerto.gob.mx/

Akio, M., Reingold, E. M., & Shimomura, M. (1988). *Made in Japan: Akio Morita and Sony*. New York: Signet Penguin Books.

Applecom. (2015). *Applecom.* Available from: https://www.apple.com/pr/products/ipodhistory/

Arce, H. (2013). *El Universal.* Available from: http://archivo.eluniversal.com.mx/finanzas/103560.html

Baker, A., & Zahniser, S. (2007). Ethanol reshapes the corn market. *Amber Waves.*

Blank, S. (2005). The Four Steps to the Epiphany (1st ed.). Academic Press.

Calorie Control Council. (2006). *Fructoseorg.* Available from: http://www.fructose.org/facts.asp

Emio national in science and technology in food. (2004). *Pnctacommx.* Available from: http://www.pncta.com.mx/pages/pncta_investigaciones_04c.asp

Garcia, L. (2011). *Analysis of the market for sweeteners in Mexico: Results report.* Autonomous University of Chapingo.

Glowingplantcom. (2016). *Glowingplantcom.* Available from: http://www.glowingplant.com/

Gower, B. (1997). *Scientific method: An historical and philosophical introduction.* New York, NY: Routledge.

Grupoelektra. (2007). *Grupoelektracommx.* Available from: http://www.grupoelektra.com.mx/Documents/ES/Downloads/Informe_Anual_Grupo_Elektra_2007.pdf

Grupoelektra. (2009). *Grupoelektracommx.* Available from: http: //www.grupo elektra.com.mx/News/Reporte.aspx?idPdf=35&tp_doc=4&sit=IREKT

Gruposalinascom. (2000). *Gruposalinascom.* Available from: http://www.gruposalinas.com/en/italika

Lock, Philip, & Gardner. (2015). *Superforecasting: The art and science of prediction.* New York, NY: Broadway Books.

Macrotrendsnet. (2016). *Macrotrendsnet.* Available from: http://www.macrotrends.net/2532/corn-prices-historical-chart-data

Mauraya, A. (2012). *Running lean: iterate from Plan A to a plan that works.* Sebastopol, CA: O'Reilly Media.

Pebble Time. (2017). *Kickstarter.* Available from: https: //www.kickstarter.Com / projects / getpebble / pebble-time-awesome-smartwatch-no-commitments? Ref = most_funded

PhRMA. (2016). Biopharmaceutical Research Industry Profile. Washington, DC: PhRMA.

Reyna, C. (2007). *Elempresariocommx.* Available from: http://www.elempresario.com.mx/Casos/Paginas/20071017162923548.aspx

Sanchez, A. (2014). *El FInanciero.* Available from: http://www.elfinanciero.com.mx/empresas/faw-vuelve-a-mexico-con-slim-tras-fracasar-con-salinas-pliego.html

Secretary of Economics. (2012). *Automotive industry: monograph.* México SE.

Shea, T. (2013). *Steve Jobs and Apple*. New York, NY: The Rosen Publishing Group.

Sony. (2009). *Time Capsule, Vol 20: Walkman finds its way into the global vocabulary.* Available from: http://www.sony.net/SonyInfo/CorporateInfo/History/capsule/20/index.html

Staley, E. (2014). *Nick Swinmurn, Tony Hsieh, and Zappos*. New York, NY: The Rosen Publishing Group.

Stone, B. (2013). *The everything store: Jeff Bezos and the age of Amazon.* London, UK: Transworld Publishers.

The Henry Ford. (2010). *The Henry Fordcom.* Available from: https://www.thehenryford.org/explore/stories-of-innovation/what-if/henry-ford/

The Tesla Motors Team. (2016). *Reserving your Model 3.* Available from: http://www.tesla.com/blog/reserving-model-3

USDA. (2007). *Outlook for U.S. agricultural trade*. Economic Research Service.

Valdez, R. (1997). Condemned the sugar industry. *El Economista*.

Vance, A. (2015). *Elon Musk: Tesla, SpaceX, and the quest for a fantastic future*. New York, NY: HarperCollins Publishers.

Zorrilla, A. (2015). Historia de las motos Italika. *Demotosonline.com.* Retrieved from http://demotosonline.com/historia-de-las-motos-italika/

Chapter 7
Lean Operation

ABSTRACT

When introduced a new solution, be it a product, a process, or a business model, the problem for the startup is the absence of a business foundation while for the established company it is precisely its existence. As discussed, market competition occurs at the level of the business formula, while company's value capture occurs through the unique arrangement of operational capabilities, strategic resources, and established relationships with key partners of the value chain; that's what we call business foundation. If the value created by the company must be captured and multiplied, it will be necessary to design a repeatable, scalable, and automatable system based on the adjustment of the business formula with the business foundation. This chapter explores lean operation.

INTRODUCTION

The test, as in medical practice, is not whether the treatment is 'scientific' but whether the patient recovers...Administration is fundamentally a practice, although, like medicine, it uses various sciences as tools. (Drucker, 1993)

DOI: 10.4018/978-1-5225-4023-6.ch007

Figure 1.
Source: Portillo, E. (2016), The Stragile Conscious Corporate Innovation Model.

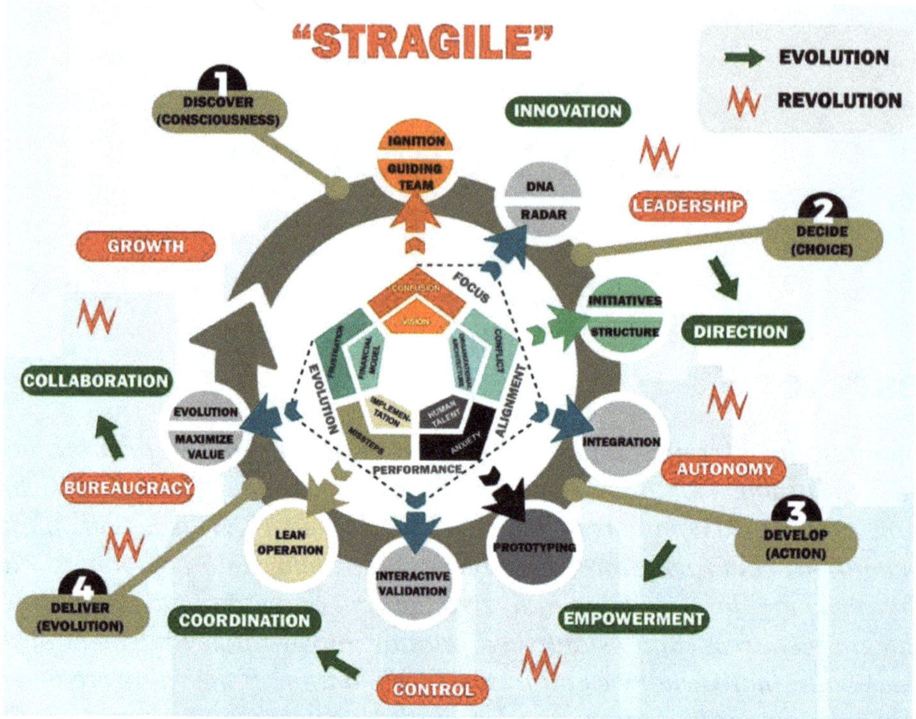

STEP 17. LEAN OPERATION

Why are there so many cases where the company that discovers a "gold mine" is not the one that "exploits" its profitably? In the mid-1970s, the Altair 8800 was introduced to the market, considered the first personal computer. Two years later, Steve Wozniak and Steve Jobs introduced the Apple I. Both created a viable minimum product for personal use computer equipment. MITS—the company that introduced and marketed the Altair 8800—as Apple Computers, ran the natural risks of experimenting with a new value proposition and developing a market, but IBM set the standard and became the dominant design of the industry. (Markides & Geroski, 2005).

For David J. Teece, capture value is a function of the nature of technology, the effectiveness of the intellectual property protection system and complementary assets. Depending on the possible arrangements between the three variables, the inventor - be it a person, a team or a complex organization - can opt for licensing or marketing. In the case of the latter, it can, in turn, integrate complementary assets under its control or ensure access through contract. As a general rule, the inventor should be able to capture the value created by his design in the possible combinations, with the exception of two cases: in both, the complementary assets are key for the reproduction of the invention and are owned by a third party. If the technical and legal appropriation is strong, the inventor must share the profit with the owners of the assets, but if the appropriation regime is weak, then it is very likely that the inventor loses to the imitators and the owners of the complementary assets. In this scenario, the recommendation is to integrate the assets under the control of the inventor or the company.

Once a design with the potential to become relevant has been validated or confirmed, the challenge of reproduction, popularly known as the scaling stage, is presented. This constitutes a "*quantum leap*" from entrepreneur to entrepreneur and from startup to enterprise, the weak point of the whole movement around invention, entrepreneurship and venture capital industry. In the case of the established company, when the new design is adjusted to the complementary assets under its control, it will be providing them with a new capacity for wealth creation; otherwise, you must develop them on your own or through a contract following the strategic criteria proposed by Teece (Teece, 1986).

Sleek Thinking and Business Systems Development

The process of iteration and validation of the business formula is often plagued by errors and waste, "*specifically any human activity that absorbs resources, but does not create value.*" (Womack & Jones, 1996). In this simple observation emerges the second characteristic of Stragile: the slender thought. The starting point of slender thought is the creation of value to identify and carry to its minimum expression all that—activities, investments, time, human effort—that in itself, does not produce.

Lean thinking defines value as the benefit provided to the client; Everything else is waste. (Womack & Jones, 1996)

The integrated network of activities (not all of these) to give birth to a successful relevant design, must be planned to improve the time and use of resources by identifying those tasks of added value and those that result in waste. It does not make sense to design a product when the existence of the problem has not been confirmed, nor to construct it when its characteristics have not been validated with the client. Likewise, it does not produce value to reproduce a design when it has not reached the adjustment between the product and the market, nor to elaborate a well-documented business plan based on assumptions and hypotheses. Doing it this way can explain many of the failures in innovation initiatives:

The sinister reality is that most new companies fail. Most new products are not successful. (Ries, 2011)

The dirty secret is that these new companies do not fail because they can not develop the product or the lack of talent of the founder. (Furr & Ahlstrom, 2011)

But because they did not discover a business formula that worked before exhausting the financial resources.

As Cooper and Vlaskovits stated:

...most entrepreneurial organizations fail because they did not develop their market, not because they did not develop their product.(Cooper & Vlaskovits, 2010)

Developing the market consists of feeding the adoption-reproduction process that characterizes innovation. In this sense, the contribution of agile thinking to the literature on organizational studies has not been valued in its right measure. The vast majority of authors have overlooked the genesis of the organization, limiting it to anecdotal references to the founders. On the contrary, Stragile makes the origin the object of study:

An entrepreneurial organization is not a small version of a great company. The entrepreneurial organization is a temporary team in search of a scalable, repeatable, and profitable business model.(Blank, Steve & Dorf, 2012)

The life cycle of the gestating company begins with the discovery stage, where the team is organized to validate the problem, the segment, and the characteristics of the solution (product); developing the first versions of the MVP until reaching the Adjustment problem-solution. In the second stage of validation, the MVP is refined and the first users are acquired; iterating or pivoting until the product-market adjustment is achieved. Once this has been achieved, the organization evolves to a third stage: Transition, to refine the business model, improve efficiency in customer acquisition and retention, adjust product reproduction processes (particularly if it is a good or physical service), and the specialization of tasks begins to take the form of an incipient formal structure. In the fourth phase, or scaling, the organization is formalized to favor accelerated growth through the acquisition of customers and the standardized reproduction of the product. From this moment on, the organization must be considered a complex system of predictable trajectory with incremental changes in the product, in the processes, and in its structure, with the only objective of capturing the value created by the market. Prior to that, there is no organization, structure, or routine system. There is simply a team with two specialized functions: market development and product development operating in an environment of high uncertainty.

In the case of established companies, the objective and context are different, but the principles of Stragile remain: Organization exists and should not be created, facing constant challenges and challenges of the environment, forcing it to change and innovate—particularly in developing new markets and new products to sustain the competitive advantage. This is not an incremental but a breakthrough innovation effort to identify, promote, and manage new areas of business other than those already known, *"taking advantage of the assets, positioning, capabilities or other resources of the established company."* (Wolcott, R.C. & Lippitz, M, 2007). The entrepreneur is one of different nature and uncertain results, seeking to be welcomed into a routine system designed for efficiency and productivity, not for innovation. In addition to validating the central hypotheses, the entrepreneural or intra-entrepreneural team must develop a model of collaboration with the routine system, taking care not to jeopardize the efficiency of the processes and the resources of the latter, and avoid being evaluated by traditional metrics and not by the advances in the milestones or criteria of the agile innovation. As Govindarajan and Trimble stated: *"...the most complex innovation initiatives and operations in progress can coexist in the same company...But their coexistence is not possible throughout it."* (Vijay, G. & Trimble, C., 2010)

The solution observed in various cases of successful innovation in established companies is the creation of a team dedicated to the development of the innovative project, and another with shared functions to perform the functions of "*bridge*" between innovation activities and efficiency activities. Having achieved the product-market adjustment and the validation of the scaling model, three things can happen: (1) The absorption of the project by an existing business unit, (2) the creation of a new business unit, or (3) the disengagement of the spin-off.

A brief reference deserves "*temporary organizations*", common in social, cultural, or research and development projects. These arise in order to solve a specific problem or produce a specific work, but once the task is completed and its object satisfied, the organization loses its *raison d'être* (purpose) and disintegrates. The nutritional problem of a marginalized community requires the intervention of diverse individual, organizational, and institutional actors. If the solution is innovative and properly implemented, the community must achieve self-sufficiency by making continuous intervention unnecessary. Only when the solution is paternalistic and supplementary can the organization remain indefinitely in return for making the community dependent and inhibiting its full development. In the cases of cultural productions, such as a film, an organization is created around the project—producers, directors, screenwriters, artists, doubles, photography, audio, montages, costumes, scenarios, to mention a few—but once the production is over, the relationship between its elements disintegrates to allow incorporation into a new project. Innovation based on technological development experiences a similar condition, referred to in the literature as "regional, national or global systems of innovation." In these three cases (and certainly others not mentioned), the validation, pivot, and iteration process of the project hypotheses remains not only legitimate but necessary, with the participation of different actors from the public, private, and social sectors.

Consequently, discovering a solution for a market—or the product-market fit—is the principle for the development of the business formula or, what the movement of agile entrepreneurship simplistically calls the escalation. The challenge for the gestating company is to create a structure with routines, methods, skills, resources, and relationships between them to create an efficient and lean system for reproduction. In contrast, the challenge for the established company is to introduce the necessary changes in each of these elements and their relationships, without affecting the profitability of other business formulas. The objective in the first case is the disruptive change. In the second; continuous improvement. In the first one the opportunity is sought, in the second the productivity. The transition from one to the other may explain many of the failed efforts of startup and innovation initiatives in the established company. Table 1 summarizes the main differences between each type of collaborative arrangement, which explains the complexity in the process of formalization of the company in gestation and that of innovation for the established company. No organization is better than the other, they are simply designed for different purposes.

Table 1.

	Company in Gestation/Intraemprendedor Equipment	Established Company
Purpose	Experimentation and discovery	Efficiency and continuous improvement
Product	In development	Standardized
Process Technology	Laboratory or workshop (flexible)	Specialized (repeatable and predictable)
Physical Assets	Generic or general purpose	Idiosyncratic and Specialized
Structure	Organic	Hierarchical
Specialization	Functional	Functional Complex or Multifunctional
Formalization	Flexible	Bureaucratic Routine

Source: Beristain, L.M. & Moreno, T. (2017) Differences between the gestating company or the intra-enterprise team, and the established company.

Difference Between Business Model, Business Formula, and Business System

Throughout the document, the concepts of business model, business formula, and business system have been referred to at different times, and it is opportune to clarify and assume a definition that can be acted upon for each of them.

As Osterwalder himself defines it, the business model *"describes the logic of how an organization creates, delivers, and captures value."* (Osterwalder, 2010). It is a representation of the business system in order to understand, visualize, or simulate the relationships among its elements. Consequently, the business system is the business model in action. A business formula, on the other hand, is the specific composition (1) of the market, (2) the business fundamentals, (3) operational capacities such as owned assets, and (4) the interaction between them. Each business formula constitutes a business system in itself and can be represented by a particular business model, so that a company has more than one business formula—or subsequently, with more than one representable business subsystem in an equal number of business models. CEMEX, the multinational company of construction materials offers concrete for the construction of avenues and highways, but also for turnkey industrial floors (Cemexmexicocom, 2012). The product can be more or less the same and use shared production, technological, and logistical resources; but the customer, the cost of acquiring different customers, and the processes of promotion, bidding, selling and administration, dissimilar, require an administrative treatment of its own and giving rise to three different formulas of business.

A company is a system. As Bertalanffy defines it, a system is *"a complex of interacting elements. Interaction means that elements (P) are in relations (R) so that the behavior of one element P in R is different from its behavior in another relation R."* (Bertalanffy, 2002) . An element consists of units of matter, energy, and information, but also on other systems, or subsystems. The arrangement of interaction involves a static-designed-for-purpose and a dynamic purpose in terms of a final state or result, according to the disposition of its processes. Each system is defined by its spatial and temporal boundaries, and the nature of the relationship with the outside, or its environment.

In a closed system, there is no communication or interaction with other systems in your environment. There is no flow of energy, matter or information, always remaining constant. On the contrary, an open system favors the permanent flow of matter, energy, or information with its environment. Complex adaptive systems, a subclass of open systems, are characterized because their elements are constantly adapting between them and their environment (Bertalanffy, 2002). One way to understand systems is in terms of existences and flows. Stocks are accumulations of elements, while flows refer to the rate of change in accumulations over time. There are input and output flows of the system, whose relationship constitutes a feedback mechanism, with delays in the communication process.

Thus, the company (the express mechanism for the reproduction of relevant designs), is a system defined externally by its purpose—to understand the mission, vision and objectives of the organization—and internally by its subsystems, components, and the interrelation between them; all to contribute to the overall purpose of the system. The limits of a complex system (established company) to innovate, is not usually in its creative or technological capacity, but in the ability to govern changes in the structure and arrangement of elements, while for the simple system (the team of the company in gestation) is to create from nothing, the system of reproduction.

The "Innate Business Instinct" for the Successful Reproduction of the Business Formula

Having an "*innate business instinct*" involves knowing the market, the customers, their habits, knowing why they buy from a specific supplier; not through modern "techniques" of management, but by "living" the concrete

Figure 2.
Source: Beristain, L.M. & Moreno, T. (2017) The system of the company: business model, business system and business formula.

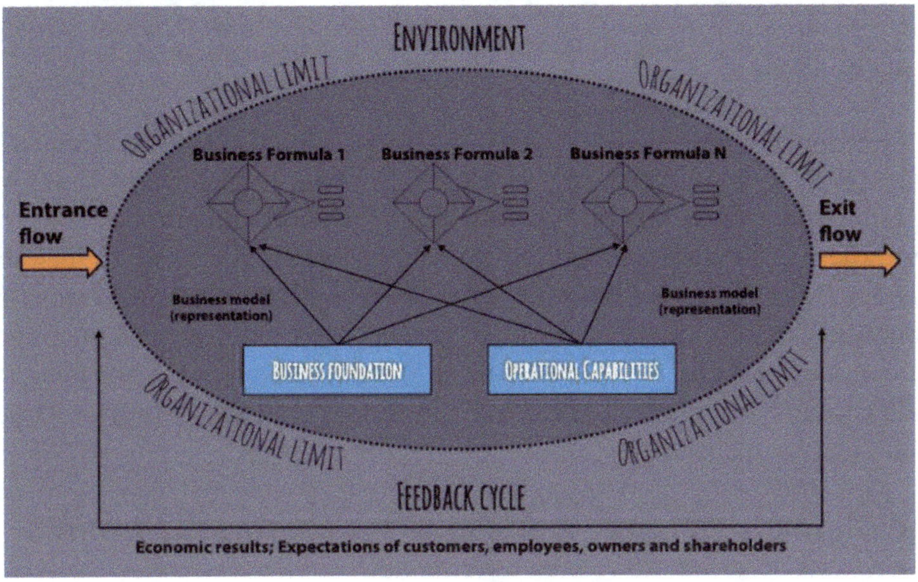

details of the business. In the author's view, the techniques are extremely valuable, but their application to specific niches, markets, and industries requires a thorough knowledge of the details of the key people—in the marketplace, supply, distribution, competitors themselves (including relevant regulatory authorities, technologies and the set of activities in which the industry in its aggregate creates or loses value). In the same way, to understand in detail the operation of the company, knowing with a certain level of dominance in which activities money is made and in which it is lost; which are unique to the company and which are common to competitors. This idiosyncratic knowledge about the features, the temperament, and the distinctive character of the company (and its market or industry), is what we call business.

Having an "*innate business instinct*" business, in addition to detailed knowledge of the important, means giving results. This is the primary function of the man of enterprise, of senior management; of management and administration at all levels. In a market economy, the results are outwardly. Having office means pursuing *better* results to precedents; not the same or worse, but better, despite market conditions. Of course, these matter, but as long as it is not a "failed state" environment, there is no excuse. As Harold Geneen stated:

It seems to me that in business there is an immutable law according to which words are words, explanations are explanations and promises are promises, but there is no reality other than results. The results and only they are the best measure of your courage, your ability and your audacity. (Geneen & Moscow, 1984)

Better results can mean different things, none exclude the other: Improved turnover, greater market share, substantial enhancements in profitability, superior cash flow, entering a new strategic market, the introduction of a new product; the improvement or innovation in the processes, to mention only some possibilities. No matter how the results are taken, they must be economic in nature; must be positive, growing and, finally, must guarantee the continuity of the company. If these three conditions are not met, the results should be questioned—decisions should preferably be challenged before putting them into action.

The Clinical Approach Applied to the Business Formula

The "*innate business instinct*" is a discipline of a clinical nature. As Peter Drucker describes it:

The test, as in medical practice, is not whether the treatment is 'scientific' but whether the patient recovers...Administration is fundamentally a practice, although, like medicine, it uses various sciences as tools. (Drucker, P., 1993)

Modern clinical practice is based on evidence; on scientific knowledge: It is observable, it can be taught and learned, is tested and its is evaluated methodically for the benefits provided. In this sense, medical knowledge such as that of business management professionals, is science. But its use and application is more of an art because of the particular situations, unique and uncertain, in which they are practiced. It is not a "*disease*", but a "*patient*" with its own existence and dignity, whose response to a treatment can be more or less average, but with the possibility of experiencing side effects, intolerance, or allergic reactions. Similarly, the business formula is unique, with unique existence and itsown history, and should be treated with the same respect that the doctor keeps for his patient. The "*quality*" of practice must be ensured by all means to avoid "*medical errors*", those arising from decision-making based on a superficial or incomplete review of the information, lack of information or lack of reliable information, the use of a wrong criterion or a false judgment, and of course, the mistake of having good faith. If the doctor is fallible, *what guarantees that the man of company or the professional of the administration is not*? Among the medical science professionals there is a maxim: *"Primum non nocere"*, translated into Spanish as "*the first thing is not to cause harm*," only then can the problem be solved. To achieve this, decisions must be the product of a detailed, thorough, and in-depth study of the single case; Of the diagnosis, the analysis practiced conscientiously to know the particular situation—even in emergency situations, and to the prudence in the decision making on the therapy or the treatment to be practiced in the concrete case. Failure to do so can have dire consequences. In this sense, clinical practice is science and art in an inseparable way. As alluded to by Joan Geneva:

Both fields of knowledge refer to life. And both, too, refer to action. After the diagnosis of the situation, both professionals [the doctor and the business manager] have to establish a therapeutic, a plan of action. And one thing or another will not always be easy.(Geneva, 1987)

Some years ago, the author approached two brothers with an unsustainable situation in the family business. Being young, the boys were well prepared and experienced, albeit outside the company and its industry. The father had suffered a health crisis that permanently disabled him, so they decided to give up their previous occupations and take over the business: A metal-mechanic company dedicated to the manufacturing of machinery for different industries. With the impetus of the youth, they dedicated themselves tirelessly to revitalize the company, introducing all kinds of professional techniques for process management, as well as control and information systems for business administration. At the heart of the activities were two areas: The production of machinery, which worked more or less in series and with standardized processes; and the workshop, where "*special orders*" were prepared, the specific modifications that Ingenious arrangements to the standard equipment and its tools. By invoicing and margin, 70% of the business was concentrated in the production of machinery. After a period of unsuccessful efforts to "*professionalize*" the workshop (accounting for about 15% of turnover), (15% of the remaining sales were provided by the maintenance services for the equipment already installed). It became evident that a change was required in the staff, particularly the chief engineer, who after thirty years serving the company, had a bad attitude toward the children's proposals and "*unwillingness to change.*" It exceeded the age for retirement and decided to retire in a generous way. Soon the problems arose. One of the main customers called the young entrepreneurs to explain that the equipment they had received did not work. Concerned, one of them went immediately to the client's floor supported by one of the technicians. The equipment worked perfectly and was built according to the specifications, but the accessories of did not "*fit*" with the existing machinery of the company. It was necessary to make modifications and this involved a design effort that would take several days, if not weeks, to do so. The customer was extremely annoyed because the production was stopped. Every day was costing a lot of money and he was letting down his customers. "*This has never happened to me in my fifteen years as your customer,*" he complained. The next day, one of the vendors complained angrily that orders for parts of their customers had not been delivered: "*The shop is not running and that puts sales at risk,*" he said. The next three days were given the task of diving into the workshop operation and discovered that the previous chief engineer was the person with all the knowledge about the needs of the customers and the parts or adjustments that each one of them required. The plans for each of these "*ingenious*" adjustments were duly documented, but it was not clear to which client the "*small details*" that were always required in the finished one were applied to. Moreover, it turned out that the company's workshop functioned as a customer design and maintenance department (there was a detailed record of the operation of each installed equipment, with scheduled parts replacement or minor adjustments, and even tentative date of replacement of the machinery in its entirety!) Another important finding was that the customer relationship was with the previous chief engineer and not the salesmen, who operated as "*pick-up orders.*" The days passed and claims accumulated. Some work was returned because it did not meet the requirements. Finally, the following rumor came: One of the main competitors of the company had made the old chief engineer a juicy offer. He was undecided, due to loyalty to his former boss (the father of entrepreneurs). Not wasting any time, after consulting with the father and with the brother's agreement, the daughter went to visit the former chief engineer to invite him as an advisor—with attractive remuneration, and partner of the company with a participation of 20%. She would dedicate herself to being the "*apprentice*" of the workshop and would allow him to organize it as he best considered it. In the end, without the operation of the workshop of that small volume of 15%, the remaining 85% would loose all its value. He gladly accepted.

This brief case serves as reference to illustrate clinical "*errors,*" even if in good faith. The best techniques of "management" and engineering—like any surgical procedure or medicine—however scientific and publicized, can not be applied without a clear understanding of the market and the company (the clients were for the value contributed from the workshop). Without an appropriate diagnosis of the pathology of a specific situation of the market or company (the basis of the business was in the knowledge of the chief engineer) and prudence in medication or therapy to implement decisions to execute—the transfer of the knowledge of the engineer "*in chief*" to the "*apprentice*" of the workshop, not to the "*general director*." It should be noted that young entrepreneurs were reluctant to acknowledge the error and accept the remediation (recipe) suggested in the situation created by them. "*Primum non nocere*." Not so, the father, who admitted that that measure had to take a long time ago. In the end, they accepted without reserve, as the resident doctor who, in the face of a patient's crisis due to a bad diagnosis or wrong treatment, runs for the experienced doctor to save both their lives; the patient and the resident.

The Desirable "Health" Status of the Business Formula

The clinical approach to business requires a clear reference to the state of health of each of the business formulas and, as a whole, the company. Unlike medical practice—more focused on monitoring the patient's health status, recovering the subject's health status, or mitigating suffering when the patient is no longer available—business practice should focus more on promotion and maintenance of the health of the company's business formulas. Of course, the first functions are usually recurrent when the company has entered into crisis, regardless of the cause and its origin—as an old professor warned, "*It does not matter if the origin of the crisis is external...the solution can only have an origin Internal.*" But then, *what is the desirable health functional state in a business formula*? Using medical terminology, *(At the end of the 1940s, David Karnofsky proposed a scale to evaluate the functional status of cancer patients. The scale ranges from 90-100 for healthy patients to 0 to declare the patient's death.),* (Karnofsky & Burchenal, 1949) *What is the asymptomatic state and full activity of a business formula and the the company as a whole*?

Plain and simple, the desired functional state is characterized by:

- A thriving market, growing and with potential to produce profits for the participating companies.
- One or more business *"fundamentals"* (Tomas, 1987) that allow the company to create a legitimate monopoly position in a niche, market, or industry, and personified in a business formula.
- A set of *"operational capacities"* that economically differentiate the company from others and allow the *"foundation"*.
- A recreatable business system—always with the same results; Scalable, with the potential to grow in volume or function to capture market value—and capable of being automated, without having to rely on a chief engineer because he can take vacations, get sick, retire from business, or die.

From the point of view of the results of the business system, as mentioned, these should be:

- Of economic nature
- Positive and increasing, not only positive (stable)
- A guarantee of the continuity of the company (sustainable)

This last condition is worth commenting on the business challenges of the moment, which is not equivalent to increase productivity or eliminate the useless costs of an operation *(A complete theory of useless costs, waste and downtime, has been most recently developed by the authors of Lean Thinking.)* to "cut" expenses. There are many entrepreneurs who have approached the author asking for help after a serious "adjustment" of the operating expenses, without realizing that they have affected the *"fundament"* or the *"activities"* of business, or both; compromising the viability and future of the company. It is equivalent to taking desperate measures for the individual who suffers from being overweight: In the end, it is not the same thing to cut a leg than to undergo a healthy regime of exercise and diet. In both cases the desired result is obtained, but obviously one of them is neither stable nor sustainable.

The economic-financial explanation indicates that, to operate, the company requires investments—known as assets: cash, banks, customers, inventories, machinery, equipment, land, deferrals, etc. These investments in turn have their sources—known as liabilities plus capital: suppliers, short-term debt, long-term debt, taxes payable, social capital, contributions, retained/lost earnings. Assets have a deadline to become sales and a profit. Liabilities have a term to be reinstated to their creditors and shareholders, and have a cost. Spending is the money used to convert investments into sales. If the term of the assets is less than the term of the most capital (P + C) liabilities, liquidity is favored increasingly and consequently there is a sustainable operation. On the contrary, if the term of the assets is greater than the term of the P + C, the suspension of payments is incurred. On the other hand, when the return on assets is greater than the cost of liabilities plus capital, there is a stable operation. Otherwise, the outcome is the bankruptcy of the company.

Finally, the metrics should not be confused with the functional state of the company, just as the symptoms should not be confused with the disease. The results are the manifestation; symptoms of the business system as a whole. In the face of the fever, many people rush to treatments based on cold water applied to ankles and forehead, bath tubs, sage tea with honey, or even the indiscriminate use of medications with acetaminophen. In the same way, entrepreneurs are faced with price reductions in the face of a fall in sales or a bank credit in the face of an unexpected reduction in cash flow, without knowing if the fever is caused by a virus or bacteria, abnormal physical activity, bone or urinary infection, and even as a first symptom (in some cases) of cancer. Before preparing a creative campaign of offers—disguised rebate of the price—, it is worth asking and answering the causes of a drop in sales, as the brothers of the metalworking company discovered.

Figure 3.
Source: Beristain, L.M. & Moreno, T. (2017) The diamond of the functional state of the business formula.

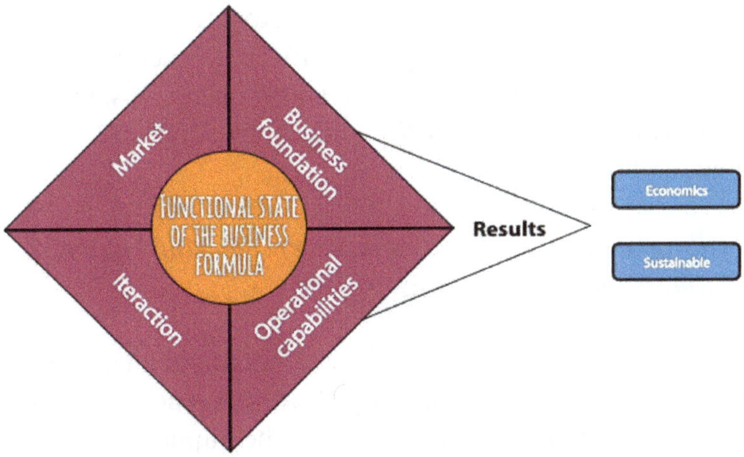

Selection of the "Right"Market

Once the frame of reference of the functional state of the business formula is established, the starting point is the customer and the aggregate of their decisions in the market. As Theodore Levitt points out clearly:

Creative marketing jumps from the obvious to the meaningful, stating that people do not buy things but solutions to problems...people who buy a quarter-inch drill do not want quarter-inch drills, they want quarter-inch holes. (Levitt, 1986)

Again, customers do not buy "*things,*" "*services,*" or "*experiences,*" only solutions to their problems. Problems of such nature that the client is willing to shed a portion of his assets to acquire a solution. There are problems with which the client can live without bothering to solve it. In other cases, it may be a "*cosmetic*" problems that would be pleasant to solve. But the problem of interest for the "*trade*" business is one that commits the customer to actively seek out the best solution within their reach and pay for it. While the problem persists, you can not fall asleep. Of course the "*thing*" (aesthetics and functionality) as the "*experience*" (service) matters, but they are not decisive in times of crisis. If the business is to solve irrelevant or cosmetic problems, you are in serious trouble, particularly when the economy goes into recession. We insist, the company does not sell products but solutions for its customers and that is the offer of value of the company. It is necessary to speak of the "*problem*" of the client and not of the needs of the client: The first is very concrete while the second can mean many things.

Consequently, in a *market economy* the customer creates value while the company creates costs. It is a mirage to think that the company "*creates or adds value.*" It transforms the inputs into a new utility, yes, but until you do not collect the sale, you do not "*realize*" the value. As Peter Druckerdefined:

The client, and he alone,

... who is willing to pay for a good or for a service, converts production into wealth and things into goods.(Drucker, 1993)

To illustrate the thesis, we present the recent case of an entrepreneur who in the first half of 1995 suffered a dry stop of sales to experience extraordinary growth towards the second half of the year. The euphoria was to become a business hangover from which, after two years, it still was not been able to recover. Production in the first six months of the fateful year only translated into an accumulation of inventories—not value, but logistical costs, machinery and equipment, space, personnel, financing, obsolescence, services, insurance, and robberies, to mention the most relevant. When sales reached historical levels in the next six months, the "*value*" was realized with the collection, but a small error of judgment was made: Historically, the company had financed the inventories under the principle of "*first entrances, first exits*". As a result, when the raw materials inventory was replenished, they were priced 60% higher than the purchase price. (In 1995 the last of the recurrent hyperinflations was experienced in Mexico, reaching approximately 52%). The company simply decapitalised. From the point of view of the trade of business and the clinical approach, it would seem obvious to proceed, first, to answer the question, *why customers are making atypical purchases* (the symptom)? As I would have to find out later, the answer was obvious (*a posteriori*, all the reasons are obvious): Because the company had not updated prices in an inflationary environment, creating a window of opportunity for customers (the cause). Second, in a situation of this nature, common sense should prevail and not accounting criteria: The inventory must be paid at the replacement price, provided it is not a deflation.

But the customer not only creates value, but also dictates, through its decisions, which products survive in the market and which products do not. As in nature, where natural selection determines the successful species (those that reproduce and multiply); in business, there is "*market selection,*" determining which products will be successful and which will disappear. If the company depends on a single product, it may also seal its destination; that is, the price of the market economy. In return, it guarantees the freedom of customers. Only as a clarification, the market economy is characterized by "*free exchange.*" If there is no such freedom, it is certainly economics, but not market.

To capture the value created by the customer, it is necessary that the value offer differs from other existing offers and is superior in solving the problem. Thus, competing products are not defined by the nature of the product or the technology used, but by its purposes—in this case, the solution of the problem. There are countless entrepreneurs who always say they have no competition, for the simple reason that there is no equal product; but there are always substitutes. After this brief discussion on the value proposition, the reader will understand clearly why the grilled chicken competes with a quesadilla or with a turkey leg cake: All struggle for a space in the stomach of the person who consumes them; the same as a spinning top, a board game, a video game, or a book compete for the customer's leisure time. From this discussion, it is clear that competition in the market is not between companies, but between products and the man of enterprise has the responsibility to decide between "*doing better*" and "*doing different.*" Not to confuse, "*doing better*" is a choice of efficiency necessary to capture market value, particularly when demand exceeds supply and when the market matures or the demand is more or less equivalent to supply. But in a market of "*perfect competition*," where supply exceeds demand, without hesitation, the choice must be for "*doing differently*."

However, there is another strategic decision of greater relevance: The choice of the market. Choosing means selecting; preferring a course of action over another, exercising the freedom to do or not and, as a consequence, voluntarily deprive yourself of the other options. Choosing the market and the clients to whom it is desired to serve is the first responsibility of the Senior Management and the axis of any business strategy. It is an act of the will that compromises, in the long term, a matter of taste, preference, and personal motivation; much as the doctor who decides to specialize in gynecology and not in angiology, or pediatrics and not in psychiatry. In the same way, the businessman selects the "*problem*" that he wants to solve through the market mechanisms. Specializing in the problem of clients in a market or industry takes years of knowledge and experience, particularly the identification and establishment of the relationship with key people along the value chain of the same and sectors of relevant support. Unfortunately, the vast majority specialize in the solution (the product or service) and not in the problem (The best exposition about the confusion between concentrating on solving the problem or concentrating on the needs of the seller (the product) is found in the multiplied article The Myopia of Marketing: Theodore Levitt. "*Marketing Myopia*", 1975). Specializing in producing and distributing telegraphs or typewriters condemns the company to a marginal and dying existence. Specializing in communications or word processing can provide a promising future. The medical oncologist does not specialize in radiation therapy; it specializes in cancer (the problem). He understands, applies, and prescribes the most appropriate treatment (the solution) for the specific case—surgery, radiotherapy, chemotherapy, or a mixture of these. He even studies and engages in basic research including clinical trials for new treatments based on hormones or transplants, looking for a new and different solution that could make the previous ones obsolete. In the same way, the businessman with "*office*" of business knows all possible solutions and exploits the most viable; one that maximizes the results of the company. At the same time, it is in the search, through small experiments of better or different ways, more valued by the customers, to solve the problem; regardless if that makes the present offer obsolete or inadequate.

How to Know That You Have Chosen the Right Market?

Whatever the specific and particular choice of the market, it must, first of all, offer the possibility of producing revenue or economic profit that results from the business. This typically occurs when:

- There is a smaller or equivalent offer to aggregate potential demand. If the total supply of the sector is in surplus, the price can approach the cost, be equal to the cost, and even lower than the cost.
- There are barriers to entry in any of its forms, limiting the entry of new competitors and with it, new supply quantities. An increase in supply alters the relationship with demand, and, of course, pushes the margin down.
- Switching costs for the customer are high.

Second, the market must have growth potential. This is experienced when:

- The annual growth of new customers is higher than the exit rate of customers. In the consumer market for disposable diapers, it is equivalent to saying that the number of newborns is greater than the number of children who stop using diapers. The inverse relationship would provoke a structural surplus of supply—if it is an industrial market, where another company is the customer, it is necessary to observe the behavior of the clients of this one.
- *Per capita* consumption of active customers is increasing or at least stable. The opposite behavior, bingo! causes an excess of supply.
- There is no change in consumer preferences. A shift towards a radically new or different supply can reduce demand, cause a supply surplus, or worse, make the latter irrelevant.
- The value of the solution for the client grows in proportion to the number of users using the same solution. In economics it is known as the "*network effect.*" A good example is Whatsapp or Snapchat as a means of communication: the more users there are with the application installed, the greater the possibilities of exploitation and value for the client.

Third, there must be a potential to diversify supply. This happens when:

- It is possible to develop a varied offer customized from a basic product, as with "*ingenious*" arrangements specific to each customer of the metalworking company referred to above, converting a standardized product into a unique value offer for the customer.
- The market for peripherals, accessories, and related products can be as important as the product itself—consider covers, loudspeakers, chargers, covers, headphones, hearing aids, handsfree and endless products that make mobile devices. Not a desirable tool, but an indispensable one as a consequence of growing demand.
- Product architecture allows you to "*assemble and disassemble*" your parts to suit different markets. The best example is computers: depending on the complexity of their structure, you can "*assemble*" a mainframe, workstation, desktop, laptop, or more recently, a Tablet (Today we know that computer equipment continues to diversify into products such as netbooks, notebooks, smartphones, and even smart watches).

In-depth knowledge of these 10 "*symptoms*" or market conditions are the kind of wisdom useful for the clinical approach to business. Some entrepreneurs are so experienced and skilled in their trade, that at the most subtle and tenuous signal can identify an opportunity or a threat, reaching to develop what we call the "business sense." That is why skepticism in those professionals of management that jump from one market to another; from one industry to another, seeking to scale positions in the corporate world by applying valuable techniques, but without context.

For these symptoms (or conditions) it is necessary to ask, *what was its nature five years ago and how is it expected to evolve over the next five? Are the conditions conducive to doing business or are they not? If they are not, is any company doing business and why?* Markets always evolve by developing their own inherent factors, particularly demographics. In other occasions, there are cycles for reasons of the environment, (particularly the economic one). Sometimes the changes are conjunctural while in others, structural. *Can it be clearly differentiated between one and the other?*

When market conditions are not conducive—there is no potential for profit, growth or diversification—it is time to explore the possibilities of "*doing different things*" in value supply, production method, management system, and even the markets served. Pretending to "do it better" in a market without potential is a certain recipe for failure.

A good example is found in Farmacias del Ahorro, a company that originated in 1991, thanks to the electronic EDI system developed by the son of a pharmaceutical merchant from the state of Chiapas with two locations in its beginning stages. By then, the traditional pharmacy was managed manually, incurring a series of inefficiencies reflected in high inventory costs, including lost sales for not having the right product at the right time. The young developer did not intend to expand the family business but to sell his *"solution"* to other drug stores. To their bad fortune (actually good because of the outcome of the story), the owners of the pharmacies of the time were not *"aware"* of the problem and consequently were not willing to pay to solve a *"cosmetic"* problem. But the system worked and produced excess positive cash flow by bringing inventory costs virtually to zero (the highest cost for a traditional business of this nature), so that, in the absence of customers for the system, the option was to expand the business leveraged in new technology and cash flow. This allowed the design of an offer of new value, different and superior to that prevailing in the commercial pharmaceutical sector. Since then, the company has grown aggressively, even in the most severe years of economic crises, just as the vast majority of companies in all sectors experience a drastic reduction in sales. By the end of 1996, the closure of traditional pharmacies had reached scandalous numbers since the beginning of the crisis, as much as 15,000 closed, according to estimates. Without a satisfactory answer to explain the phenomenon, the spokesmen of the sector went so far as to accuse Farmacias del Ahorro publicly and without any grounds of money laundering. They did not understand how one, and only one company of the pharmaceutical trade between all, could be doing business in difficult times. Farmacias Del Ahorro was not *"doing better,"* it was *"making it radically different."* The cost of inventing the new pharmacy model compared to that of the old family pharmacy was 80% lower! In this is the competitive advantage from the "office" of business: In the attributes that allow to overcome the performance of the competition in a concrete and (financially) tangible way.

The case offers two lessons: The first of these refers to the need to help the client recognize the pre-existence of a problem. In a low-potential market, they often *"ignore"* their presence—like the individual who feels perfectly healthy, turns to the doctor for a routine check-up, only to find a discouraging diagnosis. Likewise, a man with a "trade" business must go deeper in diagnosing; the *"useless"* costs of the customer by finding out how he acquires the product, employs it, installs it, provides maintenance, and eventually gets rid of it. *What value does all this give you? Are you aware of it?* And ingeniously, make the customers leaders consistent with the problem. The second refers to the task of the company and the vision of top management: To continue doing the same and pretend to do better in low or no potential markets, is folly—to hope that the conditions outside the company improve in favor, is an illusion. Albert Einstein is credited with saying, "*Madness is doing the same thing over and over again, expecting different results in each case.*" If sales decline or stagnate, when there is a substantial group of clients out there, the problem they clients have is not being adressed, but it will. In the worst-case scenario, it is being solved right now.

The Business Basis

The "right" market is, if anything, the first half of the equation. It must have te same importance to answer the question: *How does the company "captures" the value created by the customer?* With the healthy risk of being repetitive, let us remember that the client and only he, creates value; the company can only create costs in the process of turning knowledge into competencies. The "mechanism" to achieve this is what we call the business foundation, generally personified (embodied) in the value proposition. In the absence of this, any effort is useless and gains, if any, an accident. The way Farmacias del Ahorro guarantees the service to its customers (availability of the precise drug), manages the inventory, manages the operation of the chain of stores, supplies the merchandise, carries out the financial control and, in general, manages the relationship with Its customers and suppliers, would be unthinkable without the electronic data interchange system. The EDI system is the foundation; everything else is activities to make the computer system productive and profitable.

Grupo Elektra has shown exceptional ability to do business and grow through all the economic crises in its history. With a unique business system, which some describe as the new "*tienda de raya*" *(Derogatory Mexican store concept used in the Industrial Revolution placed near a "hacienda" which sold food on credit to resident laborers.)*—but unlike it, based on a free exchange and creating a purchasing power that would otherwise be unthinkable for its customers—knows and understands its market as nobody in Mexico to integrate a varied supply of goods, including white goods, appliances, electronics and accessories, photography; but also credit systems, money transfer, entertainment (TV Azteca, Total Play), communications and services, Internet (TP Link), and even motorcycles (Italika). *How do you do it*? Thanks to the in-depth knowledge of the financial capacity of the client's family to give him responsible credit—if a customer wishes to acquire a good without having the family "flow", the salesman of Grupo Elektra helps him to plan for when the personal situation is appropriate, and not as bank executives who placed mortgage loans even if they were not subject to credit; recommend offers on personal dates of importance (birthdays, anniversaries) or suggest a replacement of obsolete equipment previously purchased. Elektra is the financial advisor to its clients, the popular segment of the country (the D+, D and E segments of the population). Since its inception in 1904, they have sold on credit and know how to do it: The backlog of the national banking system in 1996 accounted for about 40% of the national GDP. At the "*worst*" moment of Grupo Elektra, past-due loans did not exceed 4.8 per every hundred. But this capability would not be scalable without DNA—the information system known as the Business Manager, developed by and for Elektra—the stores are connected to share information on credit accounts, inventory requirements, offers and prices in "real-time" sales and strict control over costs. As a result, a *"lock"* is created on the customer's term consumption *("weekly"),* whose cost of switching to competition would be extremely high (it is not subject to credit and would lose its credit history). In addition, DNA as a platform allows to assemble new goods (clothing, housing, flights, savings, health services, etc.) and diversify the purchasing power of its customers. The supply of value consists of the "*creation of purchasing power"* and the mechanism of capture, the "*lending of credit*" scaled by DNA. That is the "foundation" of business. As one company executive told the author, "*Our business is financial. In that sense, we are more like a bank and the products are only the excuse to grant the credit*" (As effectively confirmed in 2002 with the authorization of Banco Azteca).

The "*foundation"* of the business creates a *'legitimate monopoly'* in the market and guarantees the capture of the value created by the customer—in fact, manages to capture the largest proportion without losing it along the supply chain and distribution chain. Without these two characteristics, there is no "*foundation*" of business. To be sure, the "*foundation*" of business must satisfy one or more of the following conditions:

- **Unique:** The "foundation" is unique. No other mechanism of value capture in the market is similar. Such as the Telmex Telecomunicaciones infrastructure or the Banamex ATM machine network upon opening to international competition (in either case, they either use or develop this infrastructure). Or, as the unparalleled financial model in the southeast of the country, developed by the self-financing company AFISA for the acquisition of housing. This does not imply that there is no competition—in some way consumers who are not customers of these companies *"solve"* their problema—, but the supply of value that is created from these mechanisms is clearly and without confusion, differentiated as valued by the client.

- **Inimitable:** Strictly speaking, any mechanism is imitable. However, some can be simple and low cost, while others require great efforts and strong sums of investment. In an industry intensive in capital and dependent on economies of certain scale to be profitable, CEMEX has been able to develop—for over 90 years, access to quarries, an infrastructure of ovens, logistics management, control through computer systems and of own knowledge to acquire and integrate operations of cement companies, that hardly any other competitor, even with the necessary investment resources, could replicate. Today CEMEX is in the group of the 10 *"big ones"* of the world and goes for more. The same can be said for Elektra and Farmacias del Ahorro.
- **Durable:** Sooner or later, the imitation of the unique and successful mechanisms in the industry arises—the barriers to imitation are not originated in technology, but in the perception of the entrepreneurs whose best available strategy is to *"copy"* (there is nothing bad in a market economy). The key is the duration of the *"legitimate monopoly"* to capture value, position itself, and create economies of scale as economies of the learning curve—those economies that are more dependent on production history than on scale. A particular case is Aspirin, whose patent expired before 1920. The period of exclusivity allowed us to test and discover new properties and position ourselves as *"the"* product to today. Something similar happens with Gerber products for children from 0 to 2 years. Although its formulation is imitable and the packaging patent has expired, the food has the *"loyalty"* of mothers with small children. Without fear of being mistaken, most readers were fed Gerber. Loyalty is not an abstract issue, but a very specific one: Trust and confidence for a product not to endanger the health or life of the baby. At the end of the day, an adverse event is not known after it's consumption and the *"pop"* sound when opened, provides total security about the quality of the product.

- **Appropriable:** The company must be able to appropriate the value captured by the "*foundation*" of business. A case to the contrary is constituted by the entire *maquiladora* industry in Mexico, whose history since the second half of the sixties in the country and the Asian experience proves it, is only capable of capturing between 2.5 and 4.5%of the value created by customers of products assembled in plants. If you are not able to transfer knowledge and technology to develop and market your own products, it will be difficult to create wealth (the employment problem is solved, but nothing more), *(As a Mexican politician pragmatically points out, "in this matter, what matters is to secure employment for all).* Under these conditions, value is captured by design owners, patents related to technology, logistics providers and, of course, those who control marketing, but hardly captured by the "*business foundation*" based on cheap labor.
- **Non-Replaceable:** When a superior substitute arises in performance, efficiency and cost, the "*foundation*" of business ceases to be. These substitutes require time, usually decades, to emerge as the dominant mechanism, so that the entrepreneur with "*trade*" business has time to restructure and respond to change. When surrogates take a company or an industry by surprise, it usually happens by "not doing the homework." If the ability to allocate resources in inventory and location of the premises were the "*foundation*" of business of traditional drug stores, EDI system of *Farmacias del Ahorro* made it irrelevant. Similarly, the national sugar industry, dependent on the consumption of the soft drink industry, never saw the threat of high-corn fructose. Recalling that having "*business*" means to develop deep knowledge of the details, key people, technologies; the set of activities in which the industry in your aggregate creates or loses value and does the task! Of the more than fifty mills operating in the country, only one took the necessary measures—even though in the United States corn fructose had been gaining ground in the bottling industry since 1970. *Why should it not occur in Mexico*? Engineer San José de Abajo, together with the company *Procazúcar*, developed their own technology to produce liquid sugar. The problem was not the difference between the international prices between corn fructose and sugar, as all the different actors assumed, but the first was liquid and the second solid, forcing the customer (the refrigerators) to incur in additional costs to make it liquid! So advantageous was the strategy that the San José de Abajo mill reduced the cost of production, increased the volume of

business, and bought the "*quotas*" of exports from other mills to place the installed capacity in production of solid sugar.

The value-capture mechanism can take various forms. In fact, the same ones suggested by different authors of the *"resource-based enterprise approach"* (David Collis and Cynthia Montgomery are an excellent referent to familiarize themselves with this current of strategic thinking, as expressed in the article *"Competing on resources: strategy in the 1990s)"*, (Collis, D. & Montgomery, C., 2008) such as physical assets (such as TELMEX's complex telecommunication infrastructure), intangibles, patents, industrial secrets, or the know-how, to acquire and to integrate profitable companies like CEMEX), abilities, and competences (like the capacity of Elektra to grant healthy credits or of the San José de Abajo ingenuity to understand the problem and to develop technology that in both cases it talks about the supply of value) and access (to customers, as Elektra does suppliers, as *Farmacias del Ahorro* achieves, to specialized knowledge as does San José de Abajo or raw materials like It does CEMEX). Equal importance acquires the potential of "*hidden*" mechanisms, those that have them at hand, are not used to capture value. Until a few years ago, Disney characters were only used for feature films and theme parks. But from time to time, they have acquired a strategic importance in a series of products like suitcases, caps, TV shows, comics, sweets, toys of all kinds, food, promotions, mobile device protectors, USB's; the list is endless.

The curious and inquisitive reader will ask, *why differentiate between "resource" and capture mechanism*? A resource is a mean to accomplish a task. But every failed enterprise, as recent business history shows, had unproductive resources and a source of costs.

From the perspective of the inventor - whether it is a person, a team or a complex organization - the capture of value takes shape through licensing or commercialization. When it comes to the second, the possibility of capture is a function of the technical nature of the invention and the effectiveness of intellectual property rights: if it is simple to reproduce and intellectual protection is relaxed, the capture capacity is weak; On the contrary, when a knowledge is complicated to reproduce and intellectual protection is effective, then the capture capacity is strong. Any other combination supposes a moderate capacity (Teece, 2000).

The capture mechanism, by contrast, is a resource developed by the company for that purpose, transferred from a low-yielding activity to a higher-productivity one, assembled to efficiently deploy the value supply.

The sugar mills in the country have all kinds of resources, but none of them is a *"business foundation."* A capture mechanism is certainly a resource, and a very valuable one. But a resource, for the simple fact of being, is not always a mechanism of capture, particularly if it is not the source to create a *'legitimate monopoly.'*

Similarly, business fundamentals should not be confused with competitive advantage. This, as established by Michael Porter, consists of the pre-eminence of a product over competitors because of its leadership in cost or differentiation, whether in a mass market or in a niche market. In any case, the business foundation is the enabler of competitive advantage. Sometimes, the preeminence results from the control of some resource categorized as comparative advantage, meaning the geographic location, the endowment of a resource or the cheap labor. Except for the former—such as OXXO's ability to identify cost-effective locations for convenience store operations—there is no relationship to the business foundation to develop a sustainable business operation.

For the diagnosis of the "*business foundation*", the following questions should be answered: *Why do customers buy from our company and not from the competition*? *Why do competitors' customers buy from them and not from us*? *Why did this or that client lose*? And, *why did he win this time*? From the value created by the client, *what proportion does the company capture*? *What is the capture mechanism*? Defined clearly and simply. If it is not possible to do it in a simple sentence, there are problems. *What was the value captured five years ago*? *Two years ago*? If there is a difference with the value captured in the present—and it is expected to be so— *How is it explained*? Compared to the most important competitors, *what is the difference in the proportion of the value captured*? *What are the mechanisms that explain the success of the competition*? if that is the case. The answers should help you understand the nature of the business; the concrete advantages (or disadvantages) over competition, and see clearly how the value proposition solves the customer problem better. The "*foundation*" of business is something very concrete and, at the same time, requires the insight of the *"gestation of business,"* as understood by a textile entrepreneur to call the attention (energetically) to one of its employees for dealing with great deference to the client. Great volume, but without respect to the customer of a piece (cloth roll): *"Customers who buy in large quantities in the high season, make us win, but small-quantity-buying customers throughout the year pay for food; even in the off-season. Do not forget!"*

Operational Capabilities

Having the proper knowledge of the doctor and the necessary medical instruments is not the same as performing a successful surgery and save the patient's life. The first corresponds to the *"foundation"* of business and the second to the *"operational capabilities"* of the business. As Professor José Luis Tomás Lucas defined brilliantly:

Using a mining simile: The first step is to find the mine. Then, the question is how it is exploited: how to enter, excavate, make galleries, how to store material, sell it, collect, and pay.(Tomas, 1987)

In life as in business, there is an inescapable law: "*There is no such thing as a free lunch.*" In the company, many things are done and *all* of them cost. Making strategy is to make the monetary investment of those activities—or at least the most significant ones by applying Pareto principle—cost less than the competition and, at the same time, are aimed at capturing the value created by the customer. *Does this activity favors the capture of value*? If the answer is no, *can it be eliminated*? If it is not possible to do so, then you have to be creative enough to make it productive or reduce the cost to its minimum expression—as did *Farmacias del Ahorro* with inventory. From the clinical focus and the "*trade*" business, productivity means capturing helping to capture value requiring less investment, time, and cost to do so. Or as the lean business professionals propose: Eliminate the waste—everything that does not produce value for the customer or the customer would not be willing to pay for.

Operational capabilities are closely related to core competencies, as defined by Gary Hamel and C.K. Prahalad. But unlike these, operational capabilities refer to the system of tasks, skills, technologies, and resources assembled in a set of interactions, which may or may not exploit the value of these clearly-identifiable individual competencies. Consider the case of Honda Motors, known globally for its line of cars and motorcycles. It is not defined as an automotive company, but as a company with solutions for "freedom of movement", with three basic competencies: design, assembly and development of engines. These competences form part of its business foundation and at the same time enable some of its operational capabilities. But to assume that these core competencies are enough to produce an automobile or a motorcycle and bring it to market, is unthinkable. It requires a set of operations and activities organized logically around these fundamentals to be able to achieve this: that is what operational capabilities consist of. From them, they support a set of business formulas such as cars, motorcycles, but also Jets, walkers (prostheses), motor boats, robots and motors for various uses, including landmowers for home use.

The notion of operational capabilities is consistent with the concept of "*dynamic capabilities*" proposed by Teece, understood as the ability of the organization to coordinate the different tangible and intangible assets that underpin the processes of creating value, aligning them with the needs of the client (Teece & Linden, 2017). In highly competitive and changing environments, the ability of an organization to integrate, develop and reform operational capabilities as the foundation of business is essential to adjust and influence the environment.

The way *Farmacias del Ahorro* manages inventory by ensuring an exceptional level of service (measured by the number of times the customer finds the right product at the best price) or how CEMEX "*buys*" the market, economies of scale, and costs related to the learning curve are a good example. Similarly, the creation of purchasing power through the granting of healthy Elektra credit or the integration of solidarity groups of AFISA financing for the purchase of housing, are a reference of the business activities, which together give shape to the operational capacity. The organization of the work along the entire value chain of the Tepeyac de Galletas Gamesa plant or the integration of functions of all the "*partners*" (employees) in the Terciofertas fabric stores, illustrate the set of operations and owned tasks where money is earned or lost.

Business "*activities*" must be concrete and quantified, particularly the impact on the company's economy and its competitiveness in the market. It is not enough that they are anecdotal but very specific. It is not a balance sheet or an income statement, but an expression of magnitude about the impact on the business. The company Tenaris TAMSA invoiced 120,000 tons of steel tube in 1985. Ten years later, turnover reached 480,000 tons. What is surprising is not growth in sales, but sales per employee: turnover per employee went from 24 tons a year in 1985 to 218 tons in 1995 *(Information shared to the author of the note by one of the directors of the company). How did they do it*? One would think of actions like the acquisition of new technology, the installation of new equipment, or the "outsourcing" of some activities. One of the directors of the company recognized to the author's expressed question that, "*several of these actions have been taken, but they do not explain more than 30-35%. The main reason is found in the exceptional productivity achieved through the redesigning of the tasks.*" Of course there was an increase in the demand for the product, but it does not account for more than 60% of the growth in sales. The other 40%, someone else *"lost it"* to TAMSA. We can ignore the economic value of the billing and, of course, the gross margin of profit, but without a doubt, the value of the *"operational capabilities"* of this company is clearly understood.

In the competitive upholstery textile market, the only possibility for domestic manufacturers in the face of the Chinese producers' strife is flexibility in design, quantity, and quick response to customer requirements. However, the modern numerical control machinery used by Mexican producers has a restriction: the change from one color to another. The cleaning of the machinery requires an eight-hour shift, at which time they stop producing about 2,900 meters (9514.4 ft). Consequently, the minimum production lot for a color has been set at 4 thousand meters, compromising customer service in the variety of colors (an exclusive color required a minimum purchase for this quantity)—in practice, the minimum lot by color was 7,000 meters (22965.8 ft) with a deviation of up to 45,000 m (147637.8 ft) to take advantage of the arrangement. The finishing could and normally would to be different. If a customer wanted another color, it was programmed in the queue for up to 9 months! (There were more than a thousand colors in the catalog). The greater variety of colors on demand, the greater the number of changes. One of the companies in the sector, understanding the problem, was given the task of redesigning activities related to color change in the production line. Most of the modifications had little to do with "*new technology*" and certainly not "*high*" technology. One of the most successful measures was to acquire additional tools to clean the machinery and, taking care of the safety measures, to make them available to any operator along the line—previously, the tool and the materials were stored in a warehouse with a bureaucratic protocol that took from 15 to 20 minutes to receive the tools plus another similar stretch to deliver it. All of this was implemented to avoid their theft and assure proper care, (this happened every time the operator realized that it required more or different material on the cleaning process) But this measure allowed for a major change: if there was once a team of specialized personnel in cleaning for color change in the production, maintenance, and repair of machinery, now all personnel in the production line, after an extensive training effort, could do so. With other minor but non-trivial arrangements, including a negotiation with the company's union, the changeover time was reduced from eight hours to 30 minutes, allowing for greater flexibility, smaller minimum lot size, and a rapid response cycle, that no other producer (Chinese product importers included) could offer to a specialty and high cost market. Of course, installed capacity increased and with that the billing in more than 6,000 meters (19685 ft) per month (with a potential of 70,000 meters [229658.8 ft] a year). Meanwhile, the competing companies lost ground in the market.

"*Operational capacities*" are manifested as the tangible expression in the design of the product or service (the supply of value), of its ways of producing and marketing it, of the administrative and management practices in which all of the the company's tasks are organized. However, as a large majority of companies attest, not everything that is done in the company translates into business. "*Operational capabilities*" refer to the set of tasks, processes, and activities of the company in which the money is earned or lost, creating economic advantages or disadvantages compared to competing companies.

For the clinical approach to business and before so many activities carried out daily in the company, it is necessary to question—so as not to lose sight of the business sense:

- Does this activity allow you to profit substantially from the "*foundation*" of business?
- Does this activity provide gain or cost?
- How does it compare with competition? Do you win or lose money? Does it cost more or less?
- Does the process produce an economic advantage or, on the contrary, an economic disadvantage?
- Does this operation favor value capture? If not, how can I make it productive?
- If this task is not productive, why is it implemented? Should it be eliminated? If it can not be eliminated, is it feasible to acquire it from a more quality-efficient and cheaper third party?

An average company can perform anywhere from 80 to 100 different tasks. In a first effort of clinical diagnosis, it is recommended to identify from 40 to 60, the most important or obvious; analyzing the economic effect of each of them on the financial health of the company. The goal is to identify 20% of the activities that account for 80% of the results and the best way to do this is through the analysis of cash flow: cash flow is to the company like the bloodstream It is to the human body. If there is no flow, there is no viability; If it is meager, can produce a hypovolemic shock *(Also known as hemorrhagic shock).* But it is not enough to generate "flow" to suppose a healthy functional state of the company. On the contrary, it must be positive and growing. The same can be said about the profitability of the overall operation of the company (net profit to sales) and return on investment (net income to investment). Both should be positive and growing without financial *"tricks."*

To conclude this discussion, it is important to monitor the behavior of the company and its environment. Both are constantly evolving for reasons of learning, testing and error, *"accidents,"* and interaction between different agents; including the relationship with competitors, in such a way that the *"source"* of business as the importance of the *"operational capabilities"* change and move, so that attention to detail with *"clinical eye,"* becomes an inescapable task and of the utmost importance for businessman and senior management.

The Interaction Between the Elements of the Business System

One of the great tragedies of the businessman is the lack of time to enjoy life. The care of so many details in the operation gives no opportunity for a respite. In a sense, it is a modern slavery and many entrepreneurs (like executives) who, in their sixties, fall into the account they have dedicated their life to business or work, but they do not know their family or have enjoyed the product of their effort. *Is that the case*? To know, just answer a simple question: When the employer or the director of the company is absent from the business, *does the company continue to work without any setbacks*? If the answer is yes, then there is a *"business system."* If the answer is no, then the entrepreneur, the executive, or the professional *is* the *"business system."* To understand the thesis, consider a doctor, an accountant's office or a plumber's workshop. When they are absent for any reason—a business trip, vacation, or illness—the business stops invoicing. On the contrary, a good "business system" continues to invoice and can expand beyond the size of the market, even in the absence of the founder.

The company as a system, is formed by the relationship between the subsystems of the different business formulas and their "feeding" from the subsystem of the business foundation and of operational capabilities.

A business system is defined as the set of tasks and operations coordinated within the company and the relationships established with the environment. Healthy companies tend to be *"open"* systems, while those with a functional profile pathology are often *"closed."* Without going into further detail in systems' theory, healthy firms are complex adaptive systems, those whose elements (people, resources, sources, and activities) constantly adapt between themselves and their environment. In fact, it is necessary at this point to define the ostentatious concept of strategy as: the choice of specific and particular forms of relationship and linkage between the company and its environment. The environment can signify the customer market, the supplier's market, the labor market, the investment market, regulatory authorities, and of course, competing companies.

However, precision is required: the set of tasks, processes, and interaction with the elements of the system (company) must be adjusted to the reality of their environment (market). Once the above has been achieved, the system must be *"set"* in the configuration of its elements and the interaction between them, as long as the environment does not change. In other words, while an open system is required to respond to the reality of the environment, a closed system must be sought to achieve efficiency and repeatability. This is a paradox that the clinical eye of businessmen and senior management must educate and keep alert, and introduce changes in the architecture of the system when they are not yet necessary, without waiting for the reaction to a crisis.

The "correct" business system is characterized by three elements, regardless of the form it can take:

- **It Is Replicable:** Again, and again, the result is exactly the same. There is no variation. The best example is the McDonald's Big Mac. You may like it or not, but no one can deny that the taste, texture, size, shape, and temperature of the famous hamburger is always the same! Regardless of where it is consumed. That means having a *"repeatable"* system. The same can be said of the service offered by *Farmacias del Ahorro* or Elektra stores. And not to ignore, it is not limited to digital and physical products, but applicable to intangible products (services), as these companies demonstrate.
- **It Is Scalable:** System operations can grow and develop rapidly, without waning or tarnishing the result. McDonald's stores are being replicated all over the world, as we have witnessed in Mexico since the first one appeared in 1985. In contrast, the vast majority of national companies are unable to open a second location. The typical response when asking the question is: "*When? "If I can barely take care of this,"* or as many businesses presume, *"Founded in 1970 and we have no branches."* To be scalable, the system must be repeatable and automated.
- **It Is Automated:** Tasks and operations are automated when (1) they are subject to being performed by computer-programmable machines, (2) can be controlled by computer systems, or (3) are feasible to be

delegated. Virtually any human activity can be subject to these three actions, but before doing so, a simple rule of clinical business approach: never automate tasks that can be eliminated and never delegate tasks that can be automated. As simple as that. The question for activities that can be the heart of the business, the "*source*," is: *how do I automate or delegate this task*? In fact, it may be the central problem to be solved by the top management of the company—as indeed it was for the metalworking company referred to in this technical note.

The task of the businessman, the senior manager, and the entrepreneur, is to create "*business systems*" and not pretend to be a "*one-man orchestra,*" or as discussed above, the business system. Although the latter may be "heroic," it is of no use if after the 50s there are problems of the heart, divorce and estrangement from the children. Although it is not negligible to be self-employed, but even with staff in charge, there is nothing honorable in this state. The man of enterprise company, must give results and these must be better year after year, exercise after exercise, but, *at the cost of the life of the owners and the managers*?

Revisiting the Diamond of the Functional State in the Business Formula

After discussing each of the elements on the functional state of the company, the model for clinical diagnosis takes the form of a dynamic and adaptive system. One that takes "*energy and matter*" inside and outside it, constituting the motor of Interactions with their environment to create wealth. The system should never achieve "*balance.*" As John H. Holland suggests:

In fact, if a system reaches equilibrium, it does not mean that it is stable. It means it's dead. (Holland, 1996)

Thus, the functional diamond of the company, from the clinical focus of business, takes the form described in Figure 4.

At this point, the only element that requires explanation is the center of the diamond. The "*clinical approach*" of business claims to "*live*" (on a daily basis) the "*details*" of the market; the fundamentals, the activities, and the system. It does not sit in an executive office to plan and "*make decisions*" based on reports, memos, and phone calls. The clinical approach requires immersion and understanding of the "*how*" and "*why*" things are done so as not to lose the business sense. As one professor of the author used to say: *"business must be lived as the person who is married and not as the marriage counselor, that it is not the same to see the bulls from the barrier than to do it from the bullring."* It is important to talk to someone everyday and see him do his/her job; asking the reasons why a task is done in one way or another (the same is true of a customer as an operator, a supplier or an administrative). Doing it seldomly or not doing it at all would make sense if the environment and the company did not change. But evolution and changes, even the unplanned ones—particularly these—occur because of small changes in the margin, (intentional), as a result of trial and error, and even accidental. The businessman with a *"clinical approach"* has that sensitivity that leads him to develop *"business sense"; "business intuition",* which sometimes does not know to give reasons. In practice, we should suspect intuition is based on simple emotion—which, while very useful for brainstorming, is very different from the intuition developed by the *"expert judgment."* After a brief interaction with the patient, they are surprised by the accuracy of the diagnosis that the later tests confirm. The intuition of business based on the experienced judgment of who knows its customers to, its suppliers, its collaborators and its competitors face-to-face, experiences the operations of the business first-hand. And by intelligence, proceeds to validate the assumptions behind the trial, in the form of hypotheses for experimentation.

This is the essence of the clinical approach: diagnosis and action. Identify the pathology, opt for a therapy (decision), and execute it in a permanent process of feedback without borders or linear sequences. It does so in tiny steps and positive feedback loops. "*Moving away*" from the process, would seem to be two activities (diagnosis and task) occurring one after another or at best, in parallel; but when it comes to reality, it occurs in an infinite number of small, constant interactions between the two. Sometimes, it will be necessary to proceed by trial and error, but in them, it will be through controlled experiments, small scale, and low cost; as discussed in previous chapters.

The Design (or Redesign) of the Business System

Resuming the discussion of Chapters 6 and 7, once the product-market adjustment of the innovation initiative or the emergence of the dominant design has been reached, there are various arrangements for reproduction.

Figure 4.
Source: Beristain, L.M. & Moreno, T. (2017) The diamond of the functional state of the business formula.

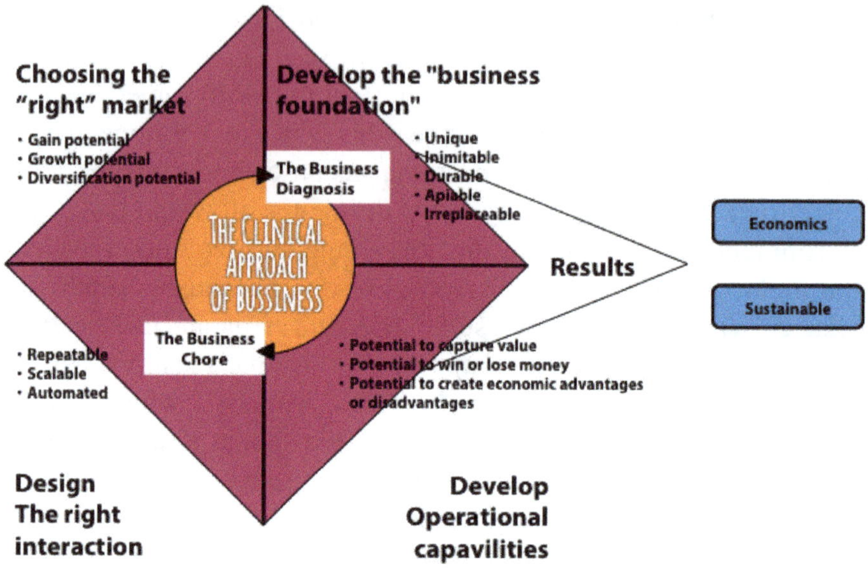

From the point of view of the company in gestation, the options are: (1) the licensing of the product, (2) the allocation of the product, (3) the association with a third party to share operational capacities, or (4) new business system. In the case of the established company, in addition to the first three, it can be chosen by (5) incorporation into the general system of the company, enriching the subsystem of business formulas or, (6) the dismemberment fo the original company—or spin off— to give life to a new system. For the purposes of this discussion, the focus is on incorporating the system of the company or the development of a new one, be it of the company in gestation or the detachment of the established company.

The starting point, as it has been emphasized from the clinical approach to business, must be the client. The one who is willing to shed part of his wealth or his credit capacity, to acquire the company's product. No matter if it is a physical or an intangible product— commonly called service—the customer has an "experience" throughout different moments of contact with the company's offer: the search for information, the possibility of experiencing the product, the purchase itself, collection, billing, installation, use or consumption, maintenance and disposal; to mention the most obvious. The businessman with a clinical approach constantly asks: *What is the best traction channel or mix of traction for the customer experience? What happens to the perception of the customer in each interaction in the chain or route of consumption*? From the customer experience perception, *how does the experience of the company compare with the experience of competing companies*? *Is it just another annoying activity*? *or is it a competitive advantage? Is the process long, costly, and bureaucratic*? *or is it agile, simple, and valuable*? *How long does it take the customer to request the product and to have it available*? The customer experience along the chain of consumption is not an accident, but the result of a design, being necessary to dedicate so much effort to the development of the product. Simply compare the experience between a traditional store like Radio Shack and a surprising Apple Store, Or between a typical e-commerce catalog like Sanborns.com.mx and Amazon.com's intelligent algorithms, or between a decent Burger King burger and the delight for parents with small children at McDonald's. If there is no accessible competitor and the product is a "*must have*" for the customer, particularly when there are shortages, this issue may be irrelevant. But even in these cases, sooner or later, the environment will be challenged where the subject is certainly strategically critical. The design of the interface with the customer must respond to the profile of the segment, remembering that it is a prerogative of the company to select the "*right market.*" Nothing is as subjective as the perception of experience and, therefore, the first strategic decision. Knowing the segment and the particular customer, with name, surname and history, as achieved by Elektra in its stores or Amazon in its digital platform, requires a strong investment in time and money to achieve an adjustment between the design of the interface and the Perception of the experience expected by the client.

Following the logic of reverse engineering, the design of the interface starts with the mapping of the desired experience for the client, providing a value superior to the offer available in the market and *always* from the perspective of the consumer. The next step is to design the process through the specific tasks and activities within the company to ensure a consistent experience. Between these, to establish reliable feedback loops and, preferably, in real time—without delays, like *Farmacias del Ahorro* does to ensure the existence of the product at the point of sale. Next, the detailed implementation plan needs to be developed and, as in experimentation for the relevant design search, to test the interface-experience fit, including the price the customer is willing to pay.

However, from the point of view of the clinical approach to business, the issue is not solved by the discovery of a relevant product or a market opportunity, but in the way of delivering and capturing the value created by the client. This being the case, the challenge for the established company is to take advantage of the possible business fundamentals—disengaging from those that are not, with the risk of incurring costs of not doing so—while for the company in gestation, it is to develop them. Even the established company may find it necessary to develop them when it decides to enter a market or a technology, or both, not related to the history of the company. The business foundation can take the form of a core competency, a strategic resource, or a strategic process. The established company is usually extremely successful when the new initiative relates to, or is adjacent to, what it already produces and has an advantage over competitors. But much of the failure of the new initiatives in the established company and the companies in gestation is explained by the inability to develop these business fundamentals. The great pharmacists have found that the threat of generics has not been such, if they themselves introduce the product into this category, taking advantage of the fundamentals developed during the time they enjoyed the temporary monopoly of the patent. Of course, it is another product, another market, and other margins. *Can anyone produce and distribute acetylsalicylic acid* (Aspirin) *better than Bayer*?

Developing a business foundation is costly and time-intensive, so the decision to do so must be the product of a thorough evaluation and a long-term vision. If the differentiation time of the value supply reduces its time and the barriers of entry become irrelevant, *what makes the company unique*? The drama in the vast majority of small and medium-sized enterprises is the absence of a business foundation other than demand, similar to the tragedy of start-ups. It is not enough to discover the gold mine; you have to learn how to exploit it profitably. When Apple's first generation of personal computers came out, the company decided to control the design of virtually everything: the memory card, circuits, RAM, microprocessor, operating system, software programs, power source, and other components of the system, only to realize that IBM had appropriated the market. For its part, IBM contributed what was good, while the processor was provided by Intel and the Microsoft operating system. A hard lesson well learned when it launched the first iPod: Apple concentrated its business foundation in the design and marketing of the product and the content (music, video, Apps), allowing third parties to take charge of the development of all the hardware elements, software, and the content needed to create the experience. In a study of value captured along the entire value chain, Apple appropriates about 58.5% of the iPhone and 30% on the iPad (Kreamer, Linden & Dedrick, 2011), reproducing the iPod's business system. A business foundation provides a cost advantage or value capture over the competition. If this does not exist or is a disadvantage, it does not exist.

To make the business foundation productive, it is necessary to develop the *"operative capacities,"* those in which the entire business of the company is organized. It is not sufficient with a medical knowledge, surgical ability, instruments and equipment to practice surgery safely. It is also necessary to apply the procedure to the correct patient, in the correct anatomical site, administer anesthesia reliably and avoid pain, counteract adverse reactions during the procedure with the use of medication, efficiently communicate critical information to all involved during the procedure, have surveillance and monitoring systems during the intervention and recovery of the patient until their discharge, and have an emergency protocol if something goes wrong. In the same way, the company must design and execute the set of tasks, activities, and processes to transform the inputs into products that the customer wants and is willing to pay for, avoiding any waste in materials, work, and time. The logic and flow of activities should consider not only internal processes, but the integration of the value chain in which the company system is "inserted", as well as the interface with suppliers and the interface with customers. The design of the task should be simple; avoiding complexity with a clear focus on cost-based competitiveness and the optimization of time and resources; clarifying that the objective is not to reduce the price, but to fight for the margins. The arrangement can be on a department or process basis, but always with the ability to change from one product to another with the least amount of friction. Also, always concentrate on those tasks in which the organization has an advantage over any other company and an architecture based on the principle of minimum investment of resources and time. The best way to reduce administrative costs is to control activities based on information, such as Farmacias del Ahorro in relation to suppliers and stock management, or Grupo Elektra with Business Manager (ADN). Sometimes this can mean sharing, renting, or leverage of assets on demand, as does UBER with cars or Airbnb with rooms.

Owning the task is recommended when knowledge and specialization are strange or scarce, and when the business foundation is put at risk. *How many tasks per business formula are there? Are they all indispensable? Can any bypass be done through the information system* (feedback)? *Should the task be done?* Sometimes, the activities are done by custom and without question to their contribution to the value created by the system. A few decades ago, information systems produced valuable reports for decision making, but as a habit of managers, these were printed on paper. In one company, printing had become a bottleneck and was administered with precision to make printing efficient throughout the day. But these imposed restrictions on delivery to the customer, since only one hour a day were remissions and invoices printed. The solution proposed by the person in charge of the area of computer science was to buy another printer and to make the necessary adjustments for its installation—the printers of great capacity had a dimension of approximately 280 x 170 centimeters, and an approximate weight of 40 kilograms. But the owner of the company, with business sense, asked for the other impressions, resulting that they were reports of sales, accounting, production, etc. With a practical sense, he gave the instruction not to print and wait tosee what happens. The first two days, no one missed the reports. On the third day, there was a person asking about the sales reports. The archivist was worried that his work would accumulate! Much of the work done in companies only produces costs.

Having achieved an efficient system of operations based on the development of operational capabilities suitable for the strategic purposes of the company, it must be confirmed that it is repeatable, consistent, scalable and, therefore, automated. The worst mistake, not *"failure"* of learning, is to automate a poorly designed process. Figure 5 summarizes the procedure for designing or redesigning the business system.

Figure 5.
Beristain, L.M. & Moreno, T. (2017) Design or redesign of the business system.

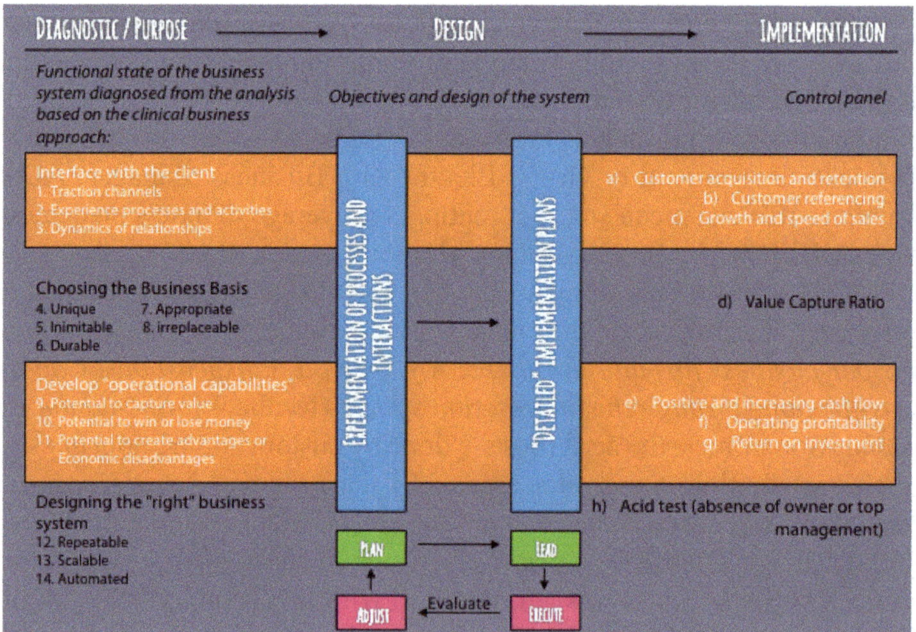

FINAL CONCLUSIONS AND RECOMMENDATIONS

There is a big difference between identifying a market opportunity with the discovery of an innovative design and being able to reproduce it profitably, as business history shows. The first social platform was not Facebook, but Sixdegrees (1997); The first MP3 music player was not iPod, but MPMan (1998); The first personal computer was not the Apple I, but the Altair 8800 (1975); The first cell phone company in Mexico was not Telcel, but Iusacell (1989). A study by Shane Snow showed that 47% of products first introduced on the market failed, while the second or third-party products had a failure rate of 8%. Part of the explanation lies in the battle for dominant design, but also part of the inability to articulate a sustainable breeding system.

For the established company, the challenge is to introduce the new product into the wealth-creation processes of the system, which involve the competition for resources of the business foundation, as well as operational capabilities with other business formulas. But sometimes, it shares the challenge with the company in gestation: to create a business system where it does not exist; developing at least a business foundation as the operative capacities to take advantage of it. The challenge of switching from a team designed for exploration to one of exploitation can be by far the difference between success and failure.

The business system must be designed to use profitably, with total business sense, and innovation; understanding the logic of the value chain for reproduction. Based on a principle of reverse engineering from the client, the activities to be performed are identified. Before each task or activity, it must always be questioned its necessity, following a decision tree: *That or that task, "must" be performed*? *Why*? If the answer is no, delete it. If the answer is yes, *is it a business foundation*? If the answer is no, *can you contract with a more efficient and cost-effective third party*? If yes, hire it. If the answer is no, *can it be automated*? When the answer is yes, do it; when the answer is no, *can you share or leverage on third-party resources*? If the answer is yes, just do it. If the answer is no, then standardize it. When the activity deals with the business foundation, then the question is, *how do I exploit this foundation*? Understanding by exploiting, gaining the most benefit or *"taking advantage"* in relation to competitors or potential competitors. Done the above, it is necessary to solve: *Is it possible to automate it*? If the answer is yes, act accordingly; Being the answer no, *how do you standardize it to make it consistent*? If you have questions about how to standardize a non-automated job, ask yourself, *how does McDonald's get products to always fit the same on trays and make the customer smile*?

REFERENCES

Bertalanffy, L. V. (2002). *General theory of the systems*. Mexico City: Fondo de Cultura Económica.

Blank, Steve, & Dorf. (2012). *The startup owner's manual: the step-by-step guide for building a great company*. Pescadero, CA: K&S Ranch Inc.

Cemexmexicocom. (2012). *Cemexmexicocom*. Available from: https://www.cemexmexico.com

Collis, D., & Montgomery, C. (2008). Competing on resources: strategy in the 1990s. *Harvard Business Review*.

Cooper, B., & Vlaskovits, P. (2010). *The Entrepreneur's Guide to Customer Development: A "Cheat Sheet" to the Four Steps to the Epiphany*. Academic Press.

Drucker, P. (1993). *Concept of the corporation*. John Day Company.

Drucker, P. F. (1993). *Management for the future: the 1990s and beyond*. Editorial Norma.

Furr, N. R., & Ahlstrom, P. (2011). Nail It then Scale It: The Entrepreneur's Guide to Creating and Managing Breakthrough Innovation. Academic Press.

Geneen, H., & Moscow, A. (1984). *High Management: The basic rules to succeed in business*. Editorial Grijalbo.

Geneva, J. (1987). *The maturation of markets and commercial strategy in business*. Editorial Continental.

Holland, T. (1996). *Hidden order: How adaptation builds complexity*. New York, NY: Perseus Book.

Karnofsky, D. A., & Burchenal, J. H. (1949). *The clinical evaluation of chemotherapeutic agents in cancer*. New York: Columbia University Press.

Kreamer, K. L., Linden, G., & Dedrick, J. (2011). *Capturing value in global networks: Apple's iPad & iPhone*. Alfred P Sloan Foundation and US National Science Foundation Paper. Available from: http://pcic.merage.uci.edu/papers/2011/value_ipad_iphone.pdf

Levitt, T. (1975). Marketing Myopia. Harvard Business Review, 53(5).

Levitt, T. (1986). *Creative Marketing*. Continental Editorial.

Markides, C. C., & Geroski, P. A. (2005). *Fast second: how smart companies bypass radical innovation to enter and dominate new markets*. San Francisco, CA: John Wiley & Sons, Inc.

Osterwalder, A. (2010). *Business Model Generation: A Handbook for Visionaries, Game Changers, and Challengers* (1st ed.). Wiley.

Ries, E. (2011). Lean Startup. Crown Publishing Group.

Teece, D. J. (1986). Profiting from technological innovation: Implications for integration, collaboration, licensing, and public policy. *Research Policy*, *15*(6), 285–305. doi:10.1016/0048-7333(86)90027-2

Teece, D. J. (2000). *Managing intellectual capital*. Oxford University Press.

Teece, D. J., & Linden, G. (2017). Business model, value capture, and the digital enterprise. *Journal of Organization Design*.

Tomas, J. L. (1987). *The business process: a useful thinking model*. Institute of Advanced Studies of the Company.

Vijay, G., & Trimble, C. (2010). *The other side of innovation: solving the execution challenge*. Boston, MA: Harvard Business Press.

Wolcott, R. C., & Lippitz, M. (2007). The four models of corporate entrepreneurship. *MIT Sloan Management Review*, *49*(1), 75–82.

Womack, J. P., & Jones, D. (1996). *Lean thinking: banish waste and create wealth in your corporation*. New York, NY: Free Press.

KEY TERMS AND DEFINITIONS

Business System: The most appropriate concept is the business model, although the latter is not enough to measure the essence of the functional state of the company. We prefer the term *system* to express the interaction between the elements of the same and their constant adaptation between them and their environment. The concept of *model* refers to the representation of a phenomenon.

Market Economy: Market economy is an arrangement where the exchange between supply and demand is free and without intention of the parties.

Chapter 8
Evolution

ABSTRACT

Once the DNA of the organization has been defined, once the culture and organizational structures required for conscious corporate innovation have been defined, once these are aligned these with the appropriate human talent, and once the processes of evaluation, control, and continuous improvement of the innovation initiatives proposed in the company have been underway, the next step is the construction of the company's evolution strategy (in some cases it can be identified as escalation, depending on the type of company or the stage of the life cycle in which it is found). This chapter focuses on proposals for a strategic approach to marketing, given that we assume that it is necessary to sell appropriately (internally and externally) the products that are worthwhile. As an important element of evolution, it is also suggested to consider issues related to revolution 4.0, specifically those related to artificial intelligence and innovation ecosystems.

STEP 19. FULL BUSINESS DEVELOPMENT AND EVOLUTION

Full Business Development and Evolution

Every story has an end, but in life, every end is just a new beginning. (Dakota Fanning, Uptown Girls, 2003)

DOI: 10.4018/978-1-5225-4023-6.ch008

Building the Paradox of the Company With an Agile Strategic Approach

In arriving at the final chapter of this exercise of questioning traditional methodologies of strategic planning, business plans, and traditional consulting and advisory models, it seems that we find contradictions at some points during the tour of the previous 7 chapters.

The truth is that, since we considered the possibility of ordering these ideas and documenting them in an orderly way, we knew that we would find successes and similarities in several of the processes that we reviewed and some of the methodologies that we combine together with those that we offer ourselves.

We knew because that is what this book is precisely about: to find models of lean innovation with a corporate approach (*Lean Corporate Innovation - ICE*).

We begin with an understanding of the new streams of slender entrepreneurship, led by Steve Blank, Erick Ries, Alexander Osterwalder, Ash Mauraya, Alex Bruton, and some other talented and visionary enthusiasts of innovation.

We understood the importance of the concepts of iteration-validation and execution-escalation. We learned the difference between building a market and building a company based on Steve Blank's models. We understood the importance of starting with the stage of problem-solution adjustment and generate pre-types initially, as proposed by Alex Bruton. We also learned to generate viable minimum products, which we could conceptualize at the stage of product-market adjustment, as proposed by Mauraya and Osterwalder in their Canvas models. We learned to differentiate the stage of construction and escalation of a company from the first two stages of adjustment, and that is at this stage when traditional management theories finally find entry and connection with the world of lean entrepreneurship.

We learned the importance of understanding and combining different methodologies and tools to improve lean corporate innovation (ICE) models. Innovative methodologies such as Design Thinking, Lean Startup, and the Business Model Canvas, mixed with some traditional innovation techniques such as QFD and TRIZ, offer truly tangible results for the benefit of the implementer.

We learned to differentiate how to apply the innovation models for companies and startups projects from those projects that require a lean corporate innovation methodology (ICE) to develop incremental, rather an

Figure 1.
Source: Moreno, T. (2017), Adapted from: Stauffer, 2011.

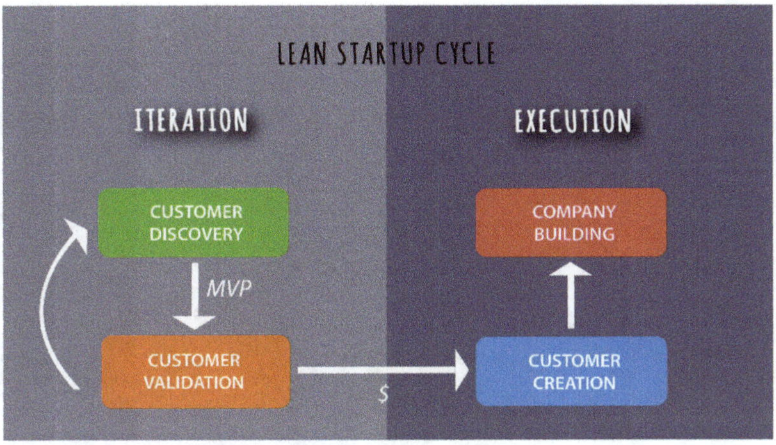

Figure 2.
Source: Moreno, T. (2017), Adapted from: Chan, 2013.

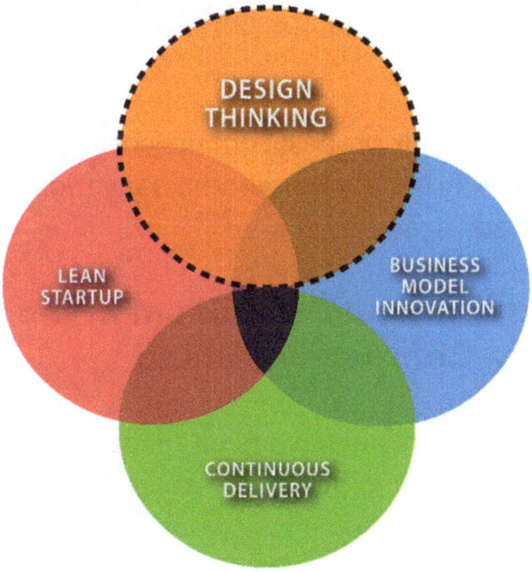

disruptive innovation strategies as fundamentally required by the initiatives in the first case.

We also learned how to manage and combine, in an orderly manner each and every one of the methodologies and tools proposed to date (Figure 3).

All of the above, leads us to understand at this stage in the writing of this book, that after having gone through the processes of creative innovation

Figure 3.
Source: Moreno, T. (2017), Adapted from: Chan, 2013.

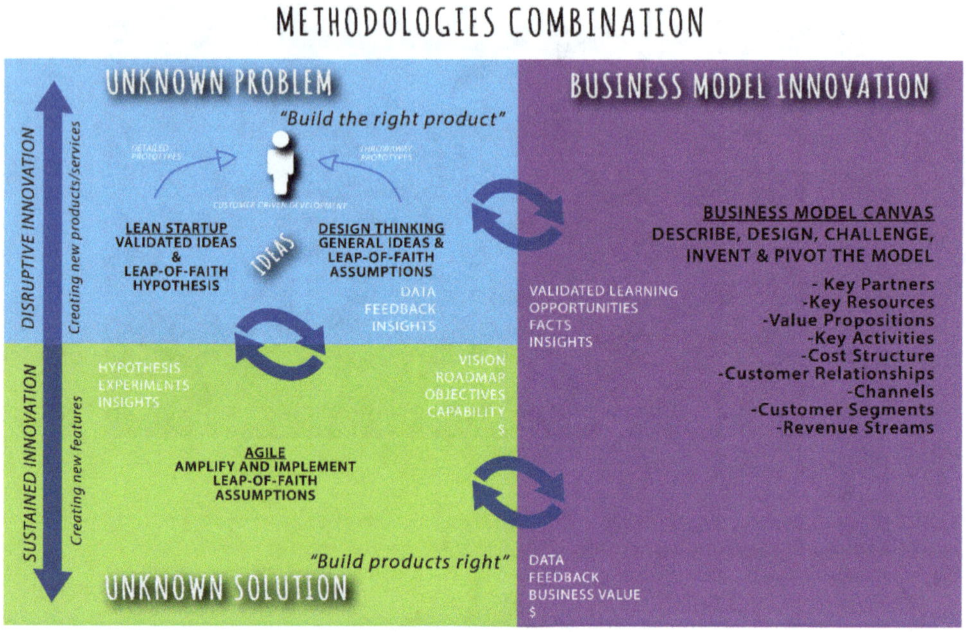

and design; and validated our proposals in the corresponding niche markets, startup projects (and spinoffs too) should be consolidated under a principle of holistic corporate construction.

The Internal and External Parallelism in the Construction of a Competitive Brand

Once the necessary adjustments have been made to massify the marketing of our ideas and proposals, it is necessary to ask the following questions to start the escalation process:

1. Does our Company have the potential market needed?
2. Do we have the appropriate solutions to solve the problems of our market niches?
3. We have the appropriate technology to meet the massive demand of our products?
4. Do we have the appropriate personnel with appropriate training for quality care in our niche market?

At this moment, we must understand that the construction of a company takes place in two parallel ways: (1) Internally (to meet all the needs of our shareholders, partners and collaborators) and (2) externally (to cover and cover each and every one of the problems and needs that our clients have, and for which they seek us.

For this, Figure 4 clearly details some internal and external aspects to consider for the correct construction of a strategy of development and organizational growth.

As we can see, the company practically has to ensure the correct linkage between three of the organizational pillars: market, employees, and technology.

First, as we have emphasized in this book, the company needs to understand, define, and live with conviction its DNA. The one that defines its purpose or reason to exist; its corporate philosophy.

Second, you need to clearly understand the problem that your niche wants to solve and generate a creative solution alternative. But once you have found your "Blue Unicorn," you must generate two key communication strategies:

Figure 4.
Source: Portillo, E. & Moreno, T. (2017), Adapted from: Rajagopal, 2009.

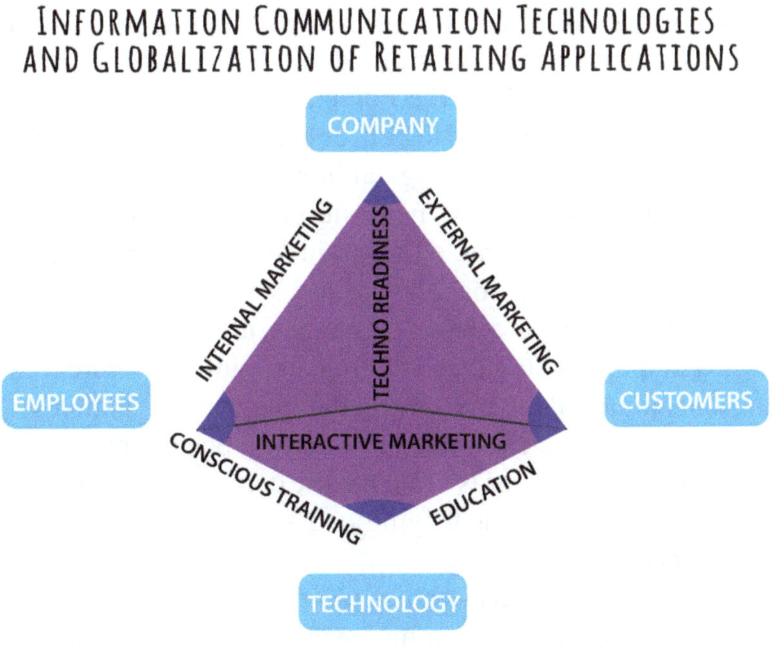

1. Attracting, convincing, and retaining the target market (external marketing).
2. Confidence and certainty in the use of technology used by the company for the benefit of its customers (educating the customer).

Finally, to accompany the customer service strategy focused on the use of technology, it must provide the staff with the training and tools necessary to interact directly and effectively with the company's customers; replacement or as a complement to automated customer service (interactive marketing).

For the internal customer (employees and collaborators), the company should develop two approaches similar to those needed to serve the external customer:

1. A strategy of persuasion and motivation for employees that allows them to understand and embrace the new technologies and processes employed. (Internal Marketing).
2. A conscious training program that inspires employees to do their best to attend to those who, as explained in the previous chapter, generate the value in the transaction (conscious training).

Finally, and not necessarily in descending order, as we mentioned earlier, in a process of escalation where the company has transcended its moments of lean operation and maximization of value, it is necessary to contemplate the development and growth of the company from a model that uses the automation of the services offered to both clients (both internal and external). Having the state-of-the-art technological infrastructure will always result in an added value and differential in the positioning of our brand.

The construction of a recognized brand in the market, is the next and last step in the Stragile methodology. There are several elements to consider when defining the value of a brand, from the most "tangible" aspects of understanding and measuring (such as financial, infrastructure, machinery, equipment, and human talent) to the most complicated (intangibles) to assess like perceptions, attitudes, intentions, and buying behaviors.

In this way, building a brand is a much more complex concept than making a minimum viable product, identifying and validating a niche market and building a company.

The construction of a brand has much more to do with the strategic aspects of growth and consolidation of a company as illustrated by Figure 5.

Figure 5.
Source: Moreno, T. (2017), Adapted from: Kottler & Keller, 2006.

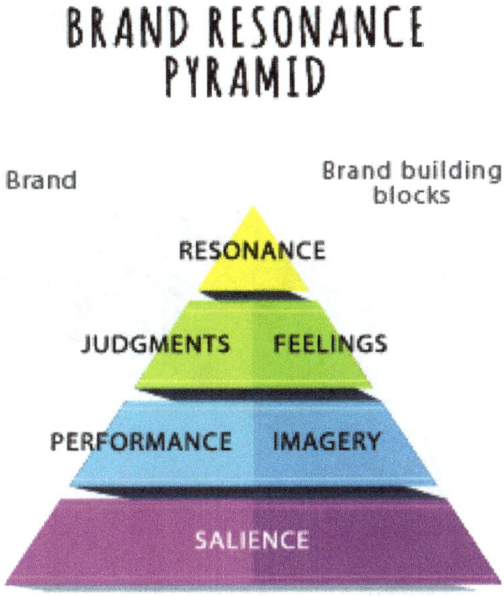

A brand must move from identity (definition of who I am), to relationships with customers (how do you and I relate?). From a communication objective focused on the information and knowledge of the brand to a communication objective focused on the reinforcement of the relationship of loyalty between the customer and the brand.

Complicated strategy in a world with more dynamism, plagued by constant appearances of highly competitive value propositions.

To achieve consolidation in the construction of brand loyalty, companies must then consider two alternative paths: internal and external.

Pioneer and Innovation Life Cycle

Let us recall the experience of our meeting with Michael Porter. The answer that he offered us to the question of whether we preferred to compete to be unique or compete to be the best, was more than illustrative. It was a lesson of strategic focus on when and why use innovation and differentiation as tools of competitiveness and commercial resilience.

In the workshops, conferences, seminars, and classes we give, we are often asked:

Figure 6.
Source: Portillo, E. & Moreno, T. (2017), Brand loyalty consolidation.

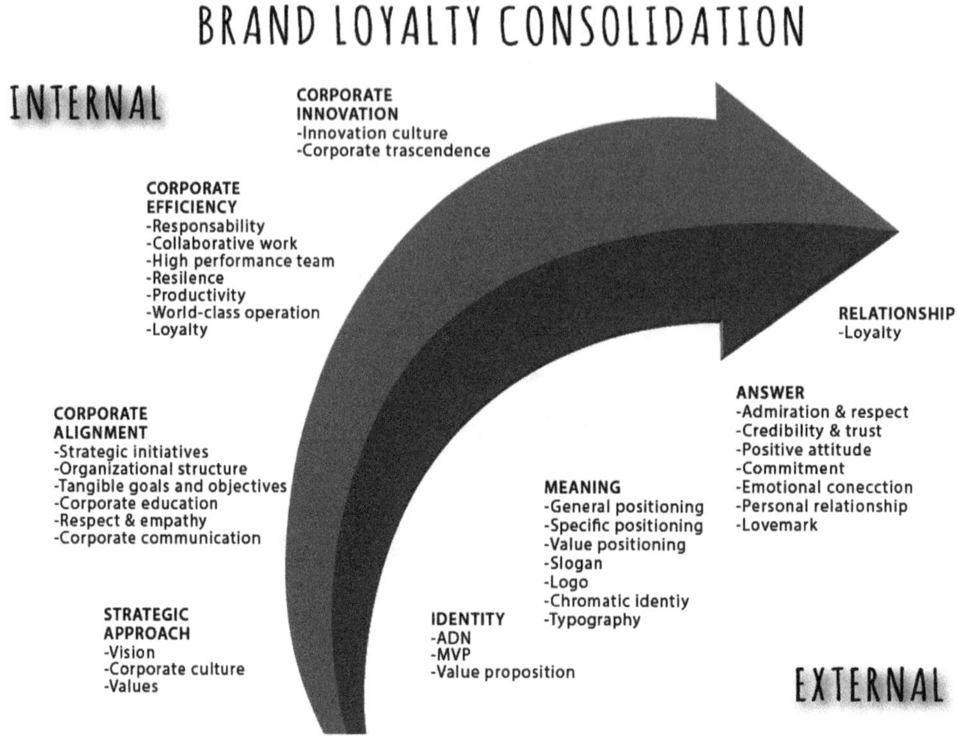

When is the best time to innovate?

The answer of course is not simple and in any case the most prudent would be to answer "It depends."

It depends on the market conditions, number of competitors, = life cycle in which the product is, = life cycle in which the company is, depends on Trump, depends on Russia, depends on North Korea, depends on NAFTA, depends on the TPP, depends on the potential new president in our country, depends on...whatever you would like to use as an argument not to explore new opportunities.

Let us see the Figure 7 under the optics of slender entrepreneurship.

In this adaptation of the Boston Consulting Group (BCG), Lex Sisney explains the lean approach required at the beginning of a project, considering market share, competition, pressure on pricing strategy management, and of course the Expected income stream (Sisney, 2012).

Figure 7.
Source: Moreno, T. (2017), Adapted from: Sisney (2016).

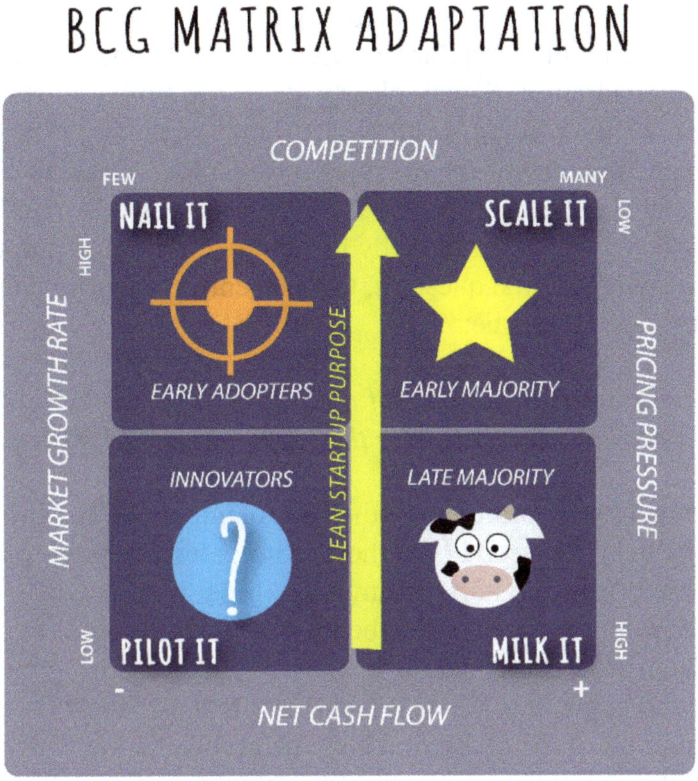

In quadrant 1 we can observe that under the Lean Startup model, this strategy is still in the "*product-market adjustment*" stage. It even uses the "*pilot for innovators*" approach.

It is clear that, given the BCG model, the gestation stage of the product does not exist. This, according to the Lean Startup Model, would be the stage of adjustment problem solution, where it moves from the pre-tendotype to the pretotype, and from the latter to the product. However, Sisney still suggests adjustments to the product at the stage of market introduction.

In quadrant 2, we can observe the idea of focusing on the priorities of early adopters. While it is true that in this phase the probability of competition is low (we would be centered in the blue ocean), it is also true that the variable "*market share*" is not yet at its highest level. The goal here is to eliminate as many objections as possible and to transform them into "*wow*" moments, where experience will turn those who "*risked*" into adopting our value proposition in our very own apostles!

Then the quadrant 3 is presented, where the objective is the mass consumption. This is where the strategy of penetration and market share are the priority. The competition begins to impact but the flow of resources also allows the financing of the variables "*delivery*" of the marketing mix.

Finally, in quadrant 4, Sisney clearly states the timing of the company. The pressure for the pricing strategy is strong, competing to wear out the competition and remove it from the market to stabilize the trend of maturity, or compete to make the last resort possible before we are the ones who leave the market when we are overtaken by the competitive value propositions.

The answer to the initial question, from an entrepreneurial point of view, I think in any case should be:

When the opportunity presents itself to create a new proposal of desirable value for the market from observing the problematic that it lives.

That is, evidently, if a product is in the stage of introduction or growth, it will hardly require to be replaced in the short term (unless the value proposal has serious questions from the start). In any case, these stages still live adjustment processes according to the needs of the market and the tools of the marketing delivery (place, promotion, and service) are at currently being implemented and extended to generate a greater penetration and participation of markets.

On the contrary, if the phase in which the product is found is in the stage of maturity, it is necessary to start at this moment the gestation stage of a new proposal or consolidation of the delivery strategy (as in the case of Coca-Cola), as illustrated in Figure 8.

The idea behind the Pioneer's Lifecycle Model is that, once someone launches into the market with a novel value proposition, sooner or later (sooner rather than later), competitors will enter with some kind of variant to compete for that "*new tribe.*"

As competitors enter, competition for market share becomes fiercer, and, as previously noted, competitive prices emerge (a situation that does not necessarily favor the end consumer, for reasons of quality).

Just before entering that competition swirl is when the pioneer, from market intelligence systems, must make the decision to continue naturally with the expected cycle of your product or detonate a strategy of product innovation that allows the re-launching of its brand.

Let us compare this cycle of life of the pioneer, with the one proposed by Sisney, where 4 preponderant forces are identified in the organization (Figure 9).

Figure 8.
Source: Portillo, E. & Moreno, T (2010), The Pioneer Lifecycle.

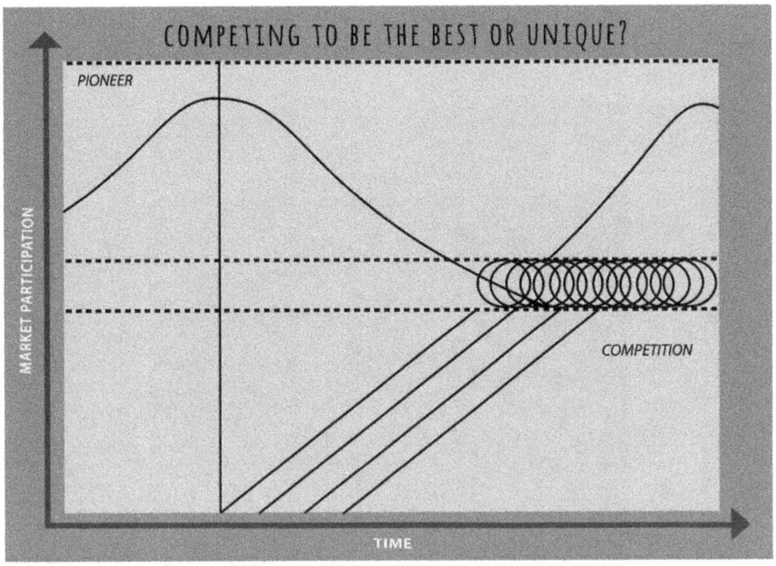

In it, Sisney defines:

... we can call out four basic patterns or forces that give rise to individual and collective behavior within an organization. They are the Producing, Stabilizing, Innovating, and Unifying (PSIU) forces. Each of these expresses itself through a particular behavior pattern. The combination of these forces causes the organization to act in a certain way. Note that, when a force is or should be dominant, it will be referenced with a capital letter.(Sisney, 2011)

For example, at birth, there is no predominant force. On the contrary, the fact that it proposes in lowercase, means that a balance between the 4 forces must be taken care of. In contrast, maturity suggests that attention should be paid to production, stabilization, and innovation, but it does not consider the concept of unification a priority (although in the case of decline it suggests not to continue investing in innovation, but to place particular emphasis on the concept of unification) in order to maintain the company's cohesion before harvesting.

As in the Pioneer Model, Sisney, 2011 notes the importance of innovation just before entering the stage of maturity, when we are entering the stage where the participation of markers increased (and consequently revenues).

Figure 9.
Source: Moreno, T. (2017), Adapted from: Sisney (2016).

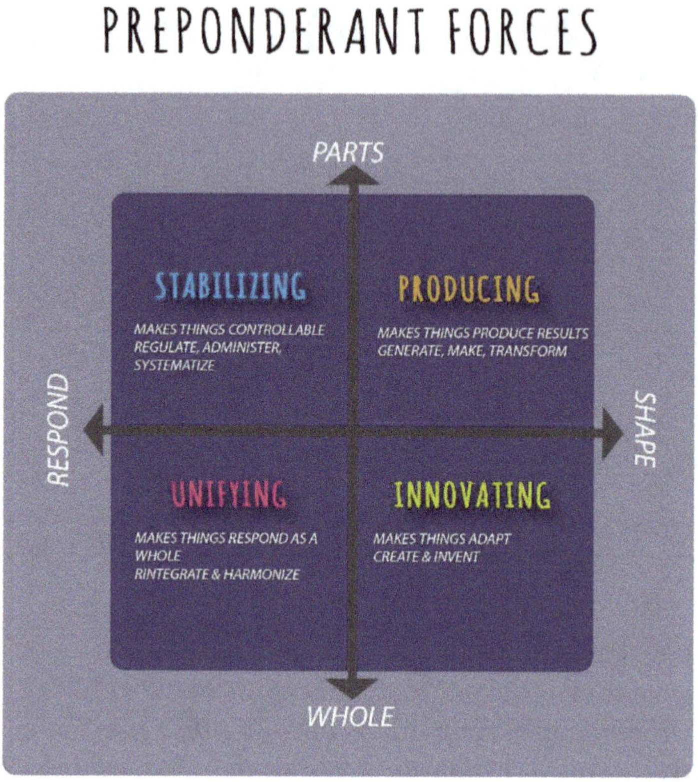

This is the moment in which the company has an extra resource to reinvent itself in what continues to improve the efficiency of its production processes from traditional quality tools and just before becoming a cost center; before reaching the limits of the Six Sigma, where innovation does not necessarily find its best ally, since the approach is quite different from Lean Corporate Innovation.

Alternatives for Growth in the Stage of Maturity

When we have entered the stage of maturity (although the reader could argue with certainty that this is not a matter of a lean corporate entrepreneurship approach at the moment), the organizational development alternatives are multiple. We simply list each and every one of the traditional books of Business Administration and Management. *Do we want to adjust the financial*

Figure 10.
Source: Moreno, T. (2017), Adapted from: Sisney (2016).

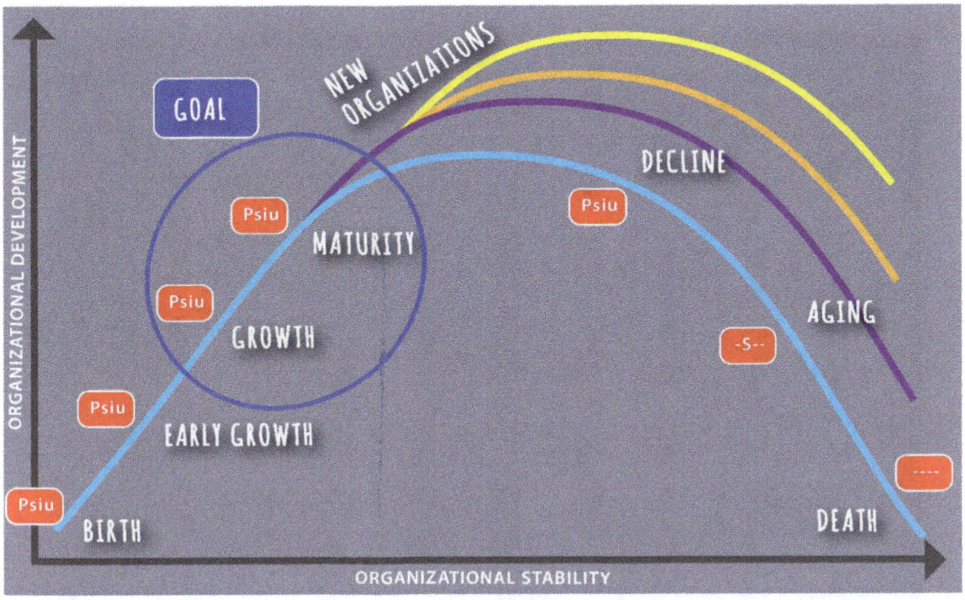

strategy? The quality of production? The development and growth of Human Talent? The Marketing Strategy? The focus and the author of your choice will surely find room at these times.

To be clear, not everything is innovation, but innovation usually finds creative solutions at all times.

However, we will continue with the marketing and sales approach, since it is the only science that attracts resources to the organization; all others represent costs and as such, are linked to a minimization approach.

Recall The Product-Market Growth Matrix by Randall L. Schultz (Schultz, 2004). Schultz's Product-Market Growth Matrix, discussed in Chapter 3.

It is very important to consider this matrix when defining and prioritizing the strategies to follow. This will help the users of the Stragile methodology in separating the variables that make up the value proposition (product and price); the variables that imply

As we noted earlier, once we enter the stage of maturity, it is necessary to capitalize the moment and start the detonation of strategies of innovation and growth.

Figure 11.
Source: Portillo, E., Chacón, J. & Moreno, T. (2016), Adaptation from: Schultz, 2004. The product market growth matrix.

1. **Market Penetration:** The effort to sell more of our current products to our customers does not require an innovative approach to our products; on the contrary, it requires that the focus of the marketing mix's "*deliverables*" is centered on the promotional strategy and services, in order to retain and attract more customers from our target market.
2. **Market Development:** When we talk about selling more of our current products to new markets we are talking about focusing on the strategy focused on the development of distribution channels or marketplaces.
3. **Product Development:** This is the strategy that is focused on the innovation of new value propositions to our current markets. It is precisely this strategy that requires the whole focus of Chapter 3 of our book to explore how to evolve such value propositions.
4. **Diversification:** Perhaps the most complex of all strategies, but also the most coherent on the basis of the dynamic evolution of markets. Complex because it requires attention to opportunities beyond what we do traditionally, requires challenging our comfort zones, challenging the

status quo, seriously questioning whether our success will be permanent or whether we can and should embark on exploring other paths. The most coherent because, precisely in the change are the most of the opportunities we seek, consistent with the DNA of the entrepreneur, the one who manages to develop his dreams, but who knows that they are not eternal and learns to implement the Stragile Cycle in every initiative that he/she undertakes.

Viewed in a different way, the product engineering and marketing approach can be represented graphically as shown in Figure 12.

This approach clearly represents the efforts that must be made according to the moment of innovation in which our products are found. It does not imply that marketing has nothing to do in the variable of product and value added; on the contrary, the idea is that it provides valuable and timely information to the engineering area to design cutting edge products that allow navigation with a large cruise on a blue ocean.

In addition, it is necessary to have a strategic clarity of the marketing approach according to the phase of the life cycle in which our products are shown in Figure 13.

This chart suggests hypothetical combinations of the strategic marketing approach. Hypothetical for a simple reason: the strategy should not be generalized and tried to implement equally to each and every one of the products and services that we offer.

Figure 12.
Source: Portillo, E. (2017) The role of Marketing and Innovation on Entrepreneurship.

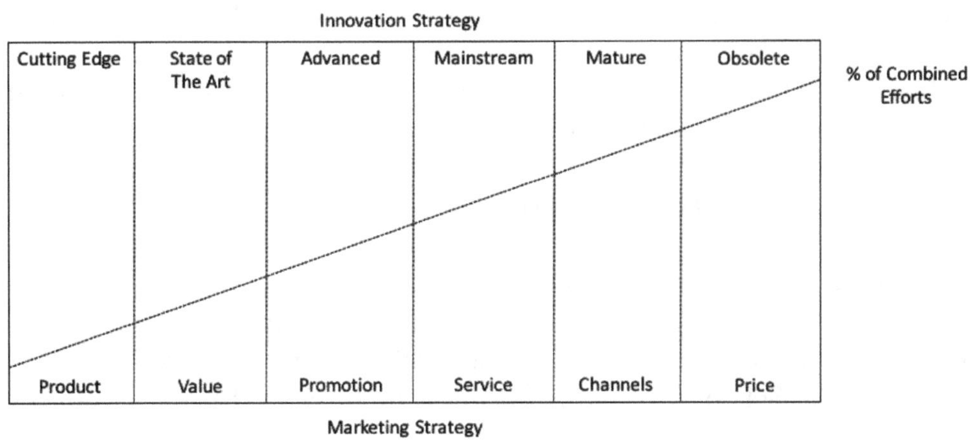

Figure 13.
Source: Portillo, E. & Moreno, T. (2017), Marketing approach according to the product lifecycle stage.

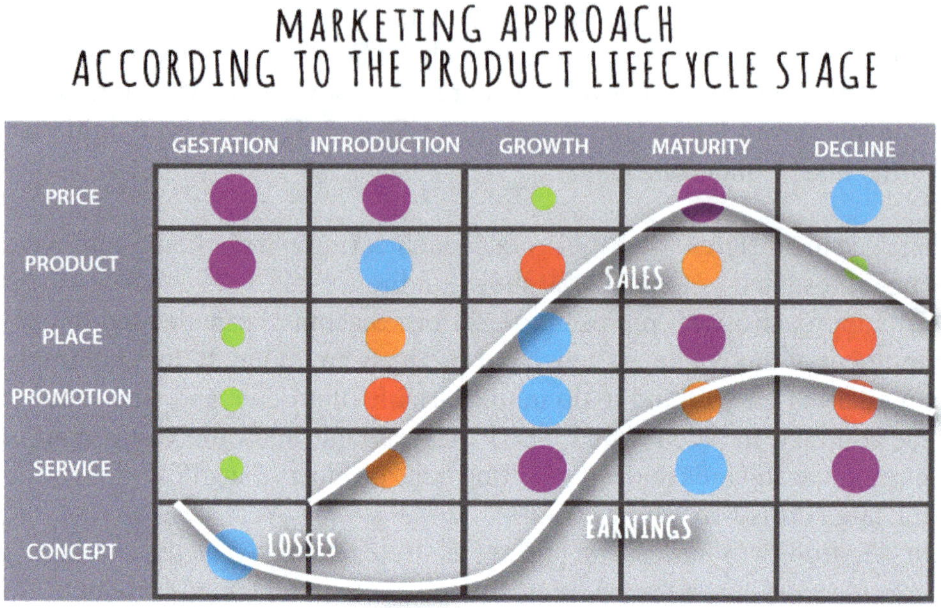

Although the general logic suggests to us, as in the role of marketing and innovation, that the approach to the product should be given at the beginning of the life cycle. It is also true that constant innovation forces us to have a permanent focus on our value propositions.

For this reason, the suggested combinations, although they may serve as a guide, must be adapted to the context of each of our value propositions.

Perhaps the only thing that can be kept constant when undertaking new projects under a lean corporate innovation approach is that the marketing mix must be separated into two phases (Table 1).

Investing in the "*delivery*" variables before validating our value proposition represents an unjustified waste of resources.

The Marketing Mix "Delivery" Variables

As we have previously noted, the purpose of this book and specifically in this chapter is not to present a summary of the marketing literature that has been developed for years from the first article presented by Professor Neil Borden of the School of Harvard Business in 1960's, where it introduces

Table 1.

Lean Startup Approach	Marketing Mix Variables
1. Value proposition development and validation	Product, Price
2. Escalation	Promotion, Place, Service

Source: Portillo, E. (2017).

Table 2.

• Product design • Price • Brand • Distribubiton channel	• Sales personnel • Advertising • Promotion • Packing	• Display • Service • Distribution • Investigation

Source: Borden, 1964. The Concept of the Marketing Mix.

the concept of "*marketing mix*" and identifies 12 important variables in the commercialization of goods and services (Broden, 1964).

Concept that in 1960 Professor Jerome Macarthy of Michigan State University reduced to 4 variables (product, price, place and promotion), (McCarthy, 1960), later recognized by the American Marketing Association as the basis of its definition of marketing in 1984, (AMAorg, 1984).

It is important to note that for the authors of this book, the fifth variable of the 4Ps (the service), is of utmost importance in today's marketing strategies. Especially from a point of retention and loyalty for the construction of a brand.

This concept will be discussed in more detail in a few paragraphs.

Let us start by suggesting some concepts related to the management of the variable promotion strategy (Figure 14).

As in the strategic approach of arketing, in the case of the variable promotion, it is necessary to contextualize its application based on different elements (such as the life cycle, market segment, competition, market share and, above all, strategic marketing objectives).

In such a way, it would be economically illogical to think (although not impossible) to invest in an intensive program of personal selling when the market does not yet have reference to our existence.

On the contrary, it is necessary that before our sales force comes to work in a personal way with the potential objections that our potential customers might have, we have previously deployed a communication exercise (preferably focused on social networks) that allows our users seek a proximity with our company.

Figure 14.
Source: Portillo, E. & Moreno, T. (2017), Promotion strategy according to the product lifecycle stage.

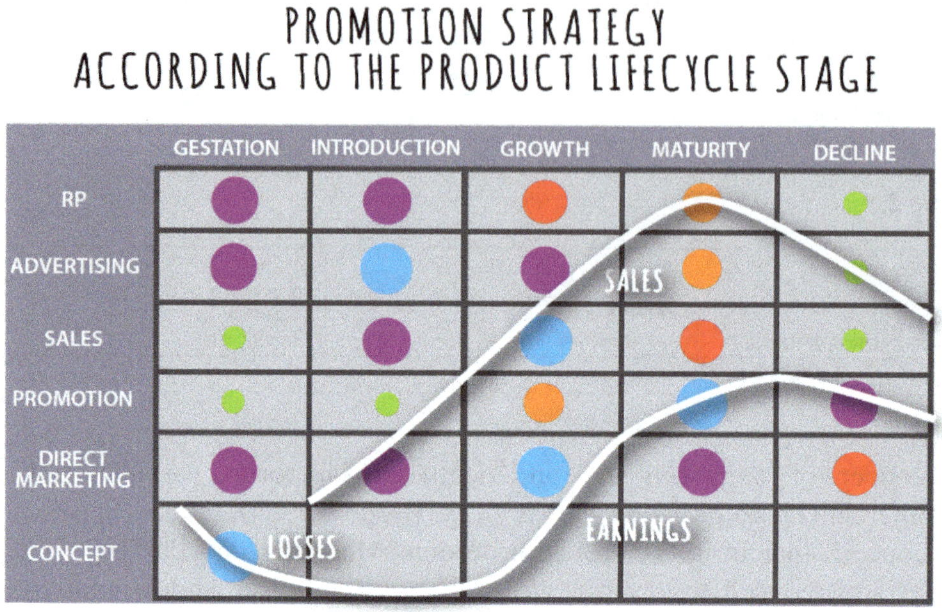

As an exercise, we also designed a table (Table 3) in which we can combine the promotional tools according to the sales process stage in which our products hover at any given time.

So, if the stage we are in is prospecting, maybe the right thing would be to use a combination of direct marketing and advertising instead of focusing on the sales force.

The variable square (distribution channels).

In relation to the distribution strategy, there are several points that we should contextualize and develop (Table 4).

Following the same approach of executive synthesis of models and theories that we have been implementing for the writing of this book, we would like to highlight one of the most recognized models, the one proposed by Professor Philip Cateora in his book *International Marketing, 1971*.

The model initially poses the complexity of planning the distribution channels under the exclusive optics of the participants in themselves in an international context.

Table 3.

Estrategia de Profesionalización de la Fuerza de Ventas								Justificación de la propuesta
Mezcla Promocional	Etapas del Proceso de Ventas							
	1.Prospectación/ calificación	2.Pre-acercamiento	3.Acercamiento	4.Presentación del mensaje de ventas	5.Manejo de objeciones	6.Cierre de ventas	7.Servicio post venta	
Publicidad								
Impresos, tv, radio								
Empaque								
Inserciones								
Películas								
Folletos								
Posters								
Carteleras								
Espectaculares								
En punto de venta								
Audiovisuales								
Promoción de ventas								
Concursos, juegos, loterías								
Premios, regalos								
Muestras								
Ferias, expos								
Demostraciones								
Cupones								
Rebajas								
Financiamientos								
Entretenimiento								
Ventas internas								
Relaciones Públicas								
Comunicados de prensa								
Discursos								
Seminarios								
Reportes anuales								
Donaciones								
Patrocinios								
Relaciones comunitarias								
Publicaciones								
Cabildeo								
Fuerza de ventas								
Presentaciones de venta								
Reuniones								
Muestras								
Ferias								
Exposiciones								
Exhibiciones								
Programas de incentivos								
Mercadotecnia directa								
Compra electrónica								
Correo								
Tele marketing								
Catálogos								
Compras por Tv								
Fax								
email								
Correo de voz								

Source: Portillo, E. & Moreno, T. (2017), Sales force professionalization strategy.

Table 4.

• Channels • Coverage • Assortment	• Locations • Inventory	• Transportation • Logistics

However, it subsequently analyzes the relationship between the actors from the point of view of the different types of relationships that must be considered when operating relations in the distribution chains.

The model highlights 5 flows to consider: The physical flow of products, the management of title deeds, payment, information, and promotion.

However, taking into account previous approaches, some other flows such as logistics, transport, customs, product insurance, and many more related to the definitions of INCOTERMS suggested in international trade, should be pointed out.

Figure 15.
Source: Moreno, T. (2017), Adapted from: Cateora, Gilly, M & Graham, 1971.

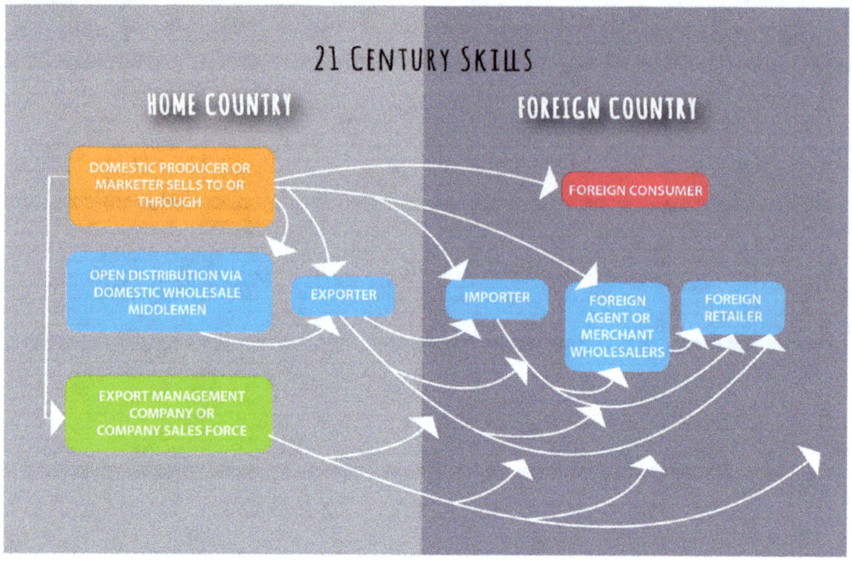

Figure 16.
Source: Moreno, T. (2017), Adapted from: Cateora, Gilly, M & Graham, 1971.

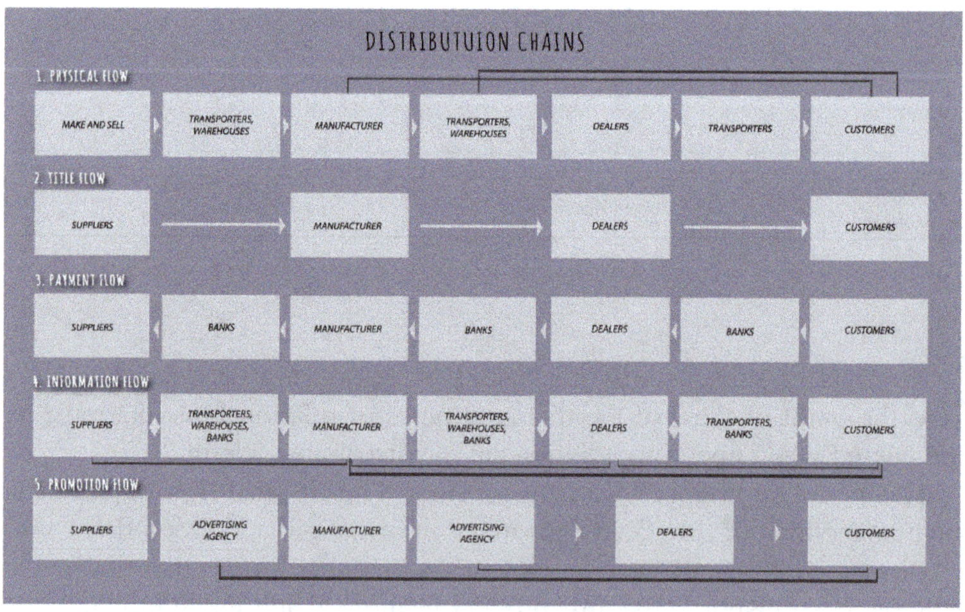

The Current Trend of "Marketplace vs. Marketspace"

Today the internet, the automation of services, and the technological infrastructure of online marketing of goods and services has come to revolutionize the sales strategies of companies. The closure of stores in some of the major commercial chains in the world is causing controversy and even served as a political platform in past elections in the United States. It was alleged at the time, that it was the Latinos who were "*stealing*" the jobs of the Americans, but there were not that many, the phenomenon to this day is quite debatable.

It is true that in the past, several labor-intensive manufacturing processes moved (simply by making production processes more efficient) to Latin American and Asian countries, and it is also true that a large number of people from Mexico and Central America were moving year afer year to the United States to develop activities within agriculture and construction.

But the real reason why these chains are closing and they are losing jobs in general at the malls, is by the tendency of consumers to buy online at much better prices.

The Marketspace Vs. Marketplace phenomenon exposed in 1994 by Jeffrey F. Rayport and John Sviokla is gaining ground and is turning shopping malls into cemeteries.

The statistics in 2017 clearly indicate this trend.

To clarify the terms, let us take the following definitions:

- **Marketplace:** Marketplace is a physical location where the buyer and seller meet each other individually and share information.
- **Marketspace:** Marketspace is an information and communication technology based electronic or online exchange environment where the buyers and sellers interact and transact in a virtual environment. (Differencebetweencom, 2015)

To this day, the general strategy of marketing and the distribution channels in specific are undergoing an impressive evolution. The business models currently evade at least two variables of marketing that until a few years ago have been considered indispensable: the media advertising and the commercial chains as commercialization points. Promotion and "*plaza*" are undergoing a real metamorphosis.

Figure 17.
Source: Moreno, T. (2017), Adapted from: Rayport & Sviokla, 1994.

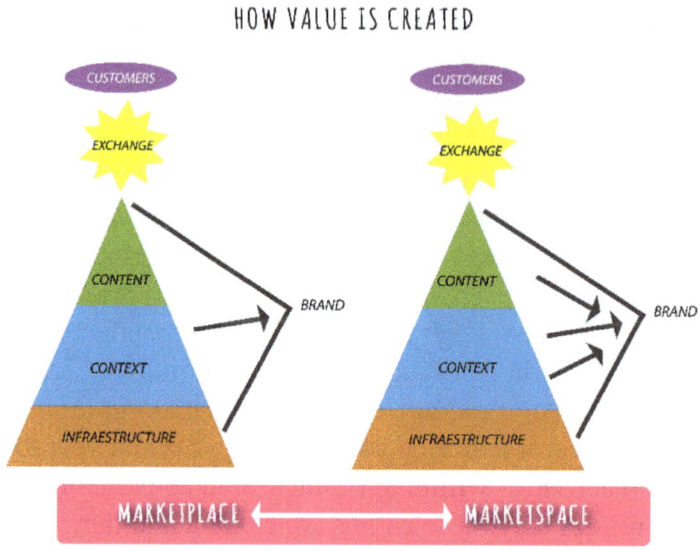

Figure 18.
Source: Moreno, T. (2017), Adapted from: Business Insider, Nudelman (2017).

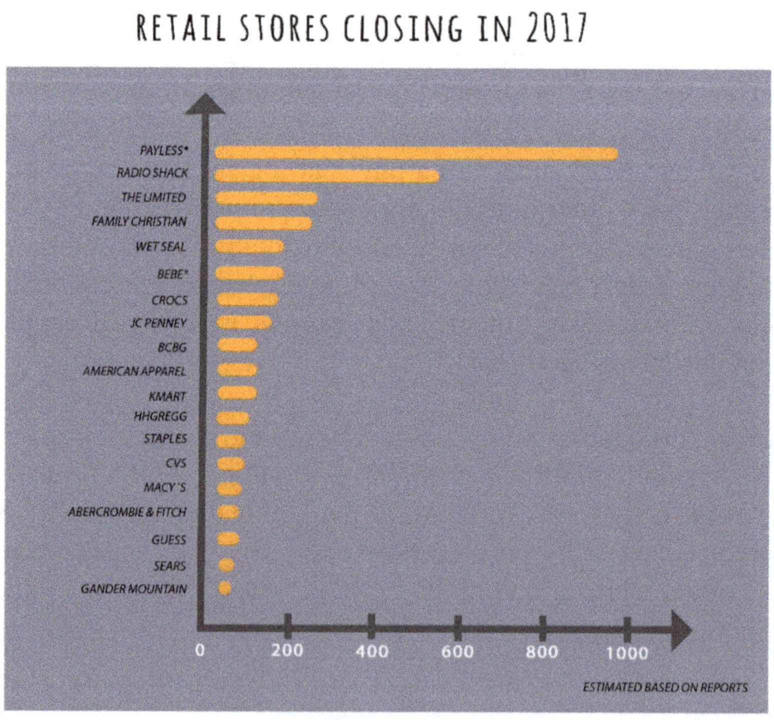

What will be interesting is to see what social gathering centers will replace the commercial plazas (in the times of our fathers the squares of the center of the city were places where people would gather, walk around, and greet each other). Will they be virtual meeting places?

What we are about to experience with the exponential growth of activity in the area of artificial intelligence is, undoubtedly, a phenomenon that will revolutionize humanity. Even the research community on the subject has urged the international community to sign a statement that suggests a code of conduct for the proper use of this technology:

We believe that AI can be a force for the good of society, but that there is also concern for inappropriate, premature, or malicious use so as to warrant the need for raising awareness of the limitations of AI and for collective action to ensure that AI is indeed used for the common good in safe, reliable, and accountable ways.. The present declaration is a step in this direction. It proposes elements towards a code of conduct for AI practitioners (both AI developers and users) in Europe. – Barcelona declaration for the proper development and usage of artificial intelligence in Europe. (Csices, 2017)

The Service Strategy

What is the reason for us to say the fifth "P" is the service? (4P's)

During my doctoral thesis, I had the opportunity to meet two great personalities from the marketing world in Chapell Hill, North Carolina: Valerie Zeithaml and Paul Bloom (coordinator of the doctoral program between UNC and ITESM).

The beginning of my thesis was guided by Valerie, the conclusion was with the support of Paul. During one of the summers we had the opportunity to be at UNC, as part of the doctoral program in which we participated, we also had the opportunity to attend a presentation Valerie made to the group.

Running into her in the classroom invited me to think about her famous SERVQUAL Model and what, in addition, brought a question to my mind: *What quality variables in the service are those that are measured when we talk about automated services*?

We are talking about 2002, when, at the time, there were very, very few articles related to automated services, and none that measured the variables related to quality in the service from self-service (or automated) technologies.

Developing the doctoral thesis in that subject, allowed me to suggest a model (Figure 19).

Figure 19.
Source: Portillo, E. & Moreno, T. (2005). Self-Service Technologies, Doctoral Thesis.

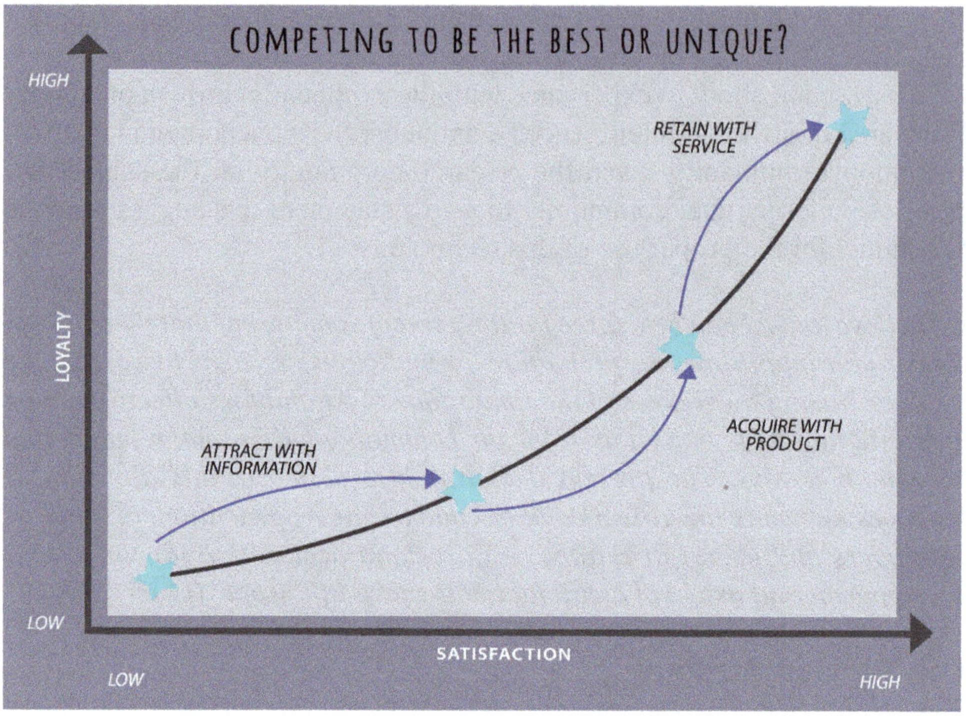

Which in turn allowed me to realize at the time of the relevance of this variable in terms of building a brand.

The service strategy is really the one that generates loyalty bonds between consumers and brands. Even behind the variables identified in the self-service technologies, we find that the personalized service still had a relevant impact on the construction of a relationship.

As explained in the chart above, communication allows us to differentiate ourselves in the sea of information existing in the market; allows us to attract the attention of our target market. A product of high quality and, above all, with a magnificent value proposition allows us to convince our market niche with respect to the reason (or reasons) why it should prefer us to the competition; however, it is the quality of human relationships that allows us to build relationships with the brand (although in terms of automated services this would have to be read differently, as will be explained below).

Consumers are looking for the convenience, ease of use, ubiquity, and efficiency of automated services, but they can not separate human attention and support when they see automated services failing. At this moment (I can not guarantee that the confidence in the technology will not be greater when the artificial intelligence surpasses the human warmth), the human being is responsible first, to give face to the failure of a machine and secondly, the only one that can offer a coherent and valuable answer that can compensate the fault. Although contrasting to what we might assume, it is not kindness what a human being seeks behind the relationship with another human being after technology fails. In fact, more than a helping hand, what we seek are timely and effective solutions that do not waste much time and add the value that we initially sought in our lives (convenience, ease of use, ubiquity, and efficiency), as we will see below.

What is the opportunity right now?

In a world of self-service, talented reps matter more than ever. But what sort of people are best equipped to handle today's customers? And how can organizations ensure that they attract and retain the most-effective reps? That's what we set out to learn. (Dixon, Ponomareff, Turner & DeLisi, 2017)

Regardless of whether service automation is gaining ground in comparison with human care, companies can not forget the importance of the role that staff plays directly behind our counter. The variables that are defined in the SERVQUAL have not changed in years, and like the good theory, will not change any time soon

The difference is that, to date, a failure in both products and services becomes viral in social networks instantly and, who has to address the flaw? Humans behind technology, off course (I insist, at least to this day, when the AI is in development).

In another Harvard Business Review article, 2017 by Matthew Dixon, Lara Ponomareff, and Scott Turner, the authors recognize at least 7 types of contributors who offer a response to our clients behind a counter (Figure 20).

The most interesting factor of this article is the surprising findings of the profiles defined by the managers as the ideal (Empathizers) with respect to the profile defined as the best performance before the customers (Controllers).

Surprisingly, Empathizers are defined as: *"service-oriented," "a good listener and communicator," and* someone who *"likes helping others."*

The definition that the managers interviewed offer of this type of collaborators is quite well justified:

Figure 20.
Source: Moreno, T. (2017), Adapted from: Dixon, Ponomareff, Turner & DeLisi (2017) Kick-ass customer service.

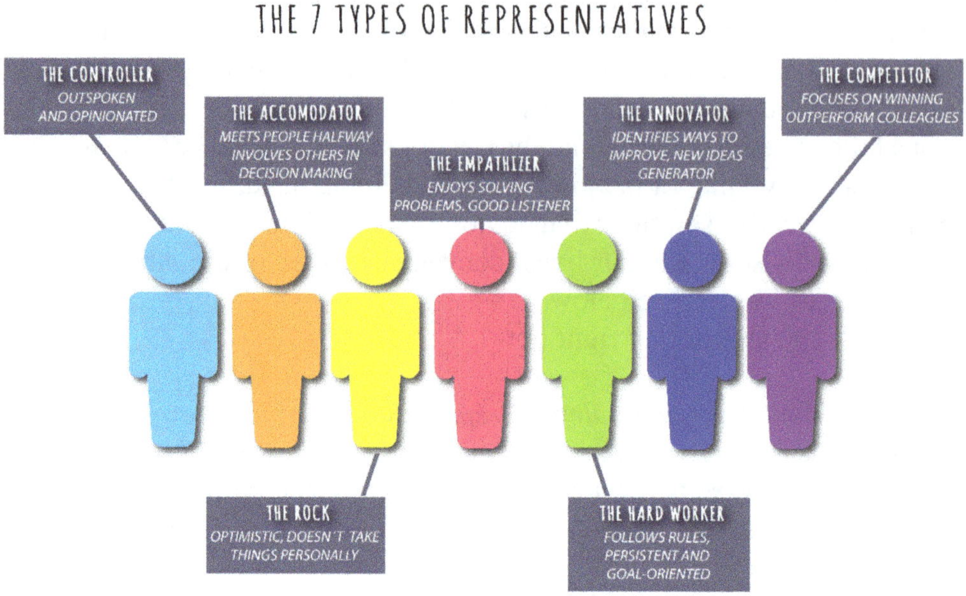

'That role is not an easy one.' Said one VP of service for a large cable operator, 'Today's customers are unbelievably impatient. As soon as we ask how we can help them, they jump down our throats. They're frustrated because of the amount of time they've had to invest on their own, frustrated by the amount of conflicting information they find on the internet, and frustrated by the thought of having to deal with a service rep. They're not calling us because they want to; they're calling us because they have no other choice.'(Dixon, Ponomareff, Turner & DeLisi, 2017)

The problem, strangely enough, is that this type of employee is not the one who performs best from the client's perspective. As it is well observed in the previous textual quotation, and as we pointed out in the introduction to this topic from my doctoral thesis, the clients look for is not to lose the time in the operation (efficiency); information that simplifies life, not that it is complicated (ease of use). Because they are guaranteed the benefit behind the operation with the automated systems (convenience) and of course ubiquity, that is to say that they can access if any problem at the moment they wish, from the place where they are and from the technological platform that they prefer.

That is, clients do not seek to make friends with people of friendly and empathetic smiles. Oddly enough, customers prefer a controller contributor profile that offers them immediate solutions, which do not take much effort to find a solution in the shortest possible time.

As the authors, themselves define it:

Why do controllers do better than their counterparts? Our structured interviews revealed that they are driven to deliver fast, easy service and are comfortable exerting their strong personalities in order to demonstrate their expertise. They describe themselves as "taking charge;" people who are more interested in building and following a plan than "going with the flow," even in social situations. They're confident decision makers, especially when nobody's in charge, and they're opinionated and vocal. As one controller explained, 'I like to take control of the situation and guide people.' (Parasuraman, Zeithhaml & Berry, 1988)

All of the above relates in terms of the performance of the service offered through self-service technologies and a trend that sooner or later will exceed the known limits of customer service today. However, in terms of human warmth, it would be convenient to refer to the basis of the SERVQUAL model (Zeithhaml, Parasuraman and Berry, 1990).

Of which we can highlight the 5 dimensions determining in relation to the perception of quality in the services:

1. **Tangible:** Appearance of physical facilities, equipment, personnel, and communication materials.
2. **Reliable:** Ability to perform the promised service dependably and accurately.
3. **Responsive:** Willingness to help customers and provide prompt service.
4. **Assuring:** Knowledge and courtesy of employees and their ability to convey trust and confidence.
5. **Empathical:** Caring, individualized attention the firm provides its customers

Comparatively speaking, we can observe certain similarities in terms of expectations of personalized service and automated services, such as tangible and reliable.

However, it is notorious that unlike personalized services, where the attitude and empathy of a human being is of great relevance, in automated services these variables are meaningless, given that the consumer is seeking more efficiency and convenience above any other thing.

However, in the construction of brands (as previously noted), human relations continue to play a preponderant role in loyalty-building.

The Lovemarks Approach suggested by Kevin Roberts in his book by the same name suggests a key distinction or differentiation in terms of building a relationship of loyalty and commitment between a brand and a client (Roberts, 2005):

This approach handles two strategic variables: Love and Respect

They are the key to differentiate and understand the strategies to follow. Roberts further identifies four quadrants or categories derived from these crossover variables. The summary that Giancarlo Giansante makes is useful for the explanation of these quadrants, as shown in Figure 21.

The four categories clearly define what we are looking for and what we do not want from our brands. We are not looking for a commodity product lacking in respect and love. We do not look for FADS with a leaky or temporary impact. We do not seek to create brands that the consumer respects, but that do not generate loyalty. The consumer understands that they are leading products and needs them, but can not find a way to connect faithfully with them.

Figure 21.
Source: Moreno, T. (2017), Adapted from: Giansante, 2015.

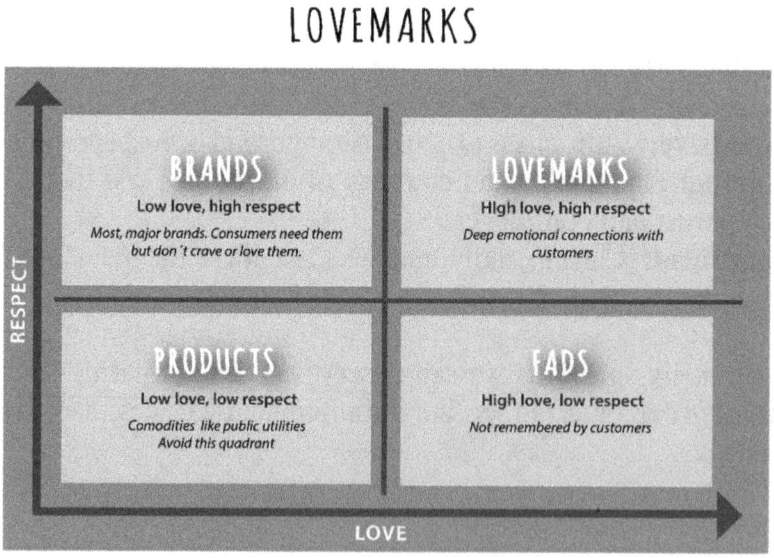

What we are looking for, in general (according to Roberts' proposal), is to generate an emotional relationship with our brands, so that the customer feels a passion for the brand; an identification and loyalty to it. Examples of this emotional level may be illustrated in the sports' world. The passion that a fan unfolds to defend the colors of his favorite team is the ultimate goal of the Lovemark concept.

As we can remember, in chapter 1 of this book we were talking about the two types of feeling or emotions that govern our lives: love and fear (with their different expressions). The interesting thing about Roberts' proposal in terms of EVOLUTION of this final chapter is that it suggests just that: To reach the next level as a brand (or as a company), it is important that love or devotion to customer service becomes an organizational culture.

Anyone who fails to experience this level of emotionality will run the risk of omitting a "*moment of truth*" and, therefore, put an entire organizational strategy at risk.

This level of evolution is not simple, it is a challenge for corporate leadership.

The previous theme implies a different conceptualization and application of the concept "*love.*" Usually, we associate it with the feeling of desire expressed between a couple, or with the feeling of appreciation and respect for parents, siblings, or friends. On many occasions, we misrepresent it when we limit it to a sexual expression.

However, in a broader sense, love means voluntary surrender to the well-being of others. Or as Tim Sanders defines it:

I do not think there is anything bigger than love. I mean, I'm not sure there could be more than love, because love is about it all. I had a hard time giving a definition of love to write my book. In the end, I define love as the selfless will to make another person grow.(Sanders, 2011).

Perhaps this is one of the best methods to defining a quality approach to service. An approach where a human being consciously and objectively seeks to generate extraordinarily pleasant solutions for the people he/she attends.

Explanations related to positive thinking, right attitude, or use of correct verbal and non-verbal language can be found by the hundreds. However, in another recent article in the Harvard Busines Review, we find some interesting data on the chemical effect of using appropriate words in human behavior (for the case of this subject, we refer, of course, to the impact language can act in a behavior of loyalty).

Figure 22.
Source: Moreno, T. (2017), Adapted from: Glaser & Glaser, 2014.

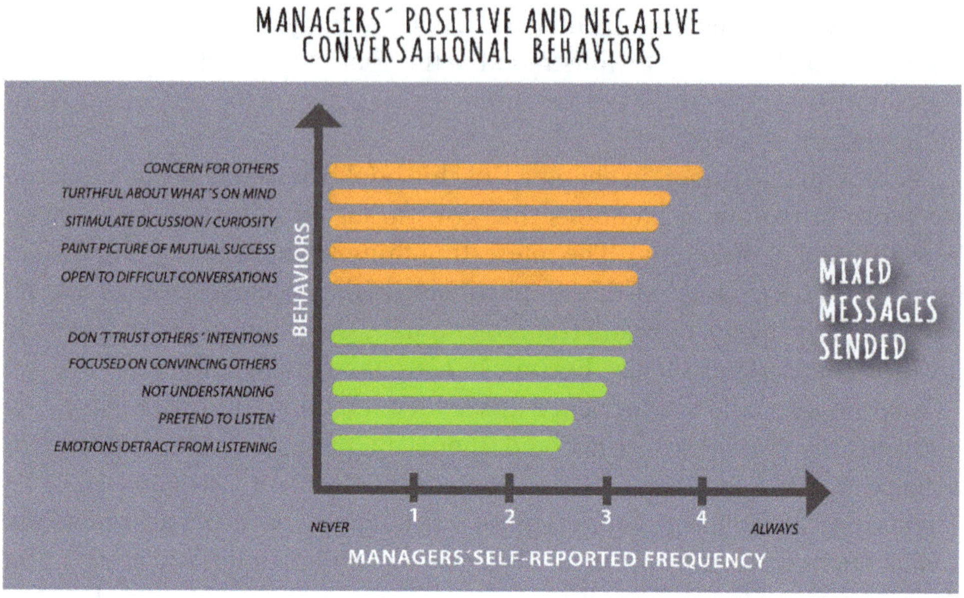

In this article, the authors suggest that positive comments tend to linger less in a person's brain than negative comments. That is, we may be receiving positive comments for several weeks or months, but a badly-made or misunderstood comment may end a relationship.

This phenomenon helps to explain the reason why it takes so long to build a relationship and how we can lose it by carelessness. The article explains that the chemical reaction our body experiences when receiving a negative comment, when experiencing a criticism, rejection, or fear due to human interaction, our body produces high loads of cortisol, a hormone that tends to paralyze rational thinking, activating our attitudes to avoid conflict, and generating protective behaviors (in some cases, fear postures tend to paralyze people when they experience fear).

This type of chemical that influences of our body, in general, leads us to magnify problems and conflicts more than they really are. To make matters worse, it is suggested that this type of effects can last up to 26 hours.

On the contrary, positive situations and conversations generate a different chemical reaction! They release oxytocin, a hormone that makes us feel good and allows us to trust and communicate better with other human beings. The bad news is that oxytocin is metabolized faster than cortisol, a situation that immediately kills oxytocin in the presence of any negative situation.

Data like these, forces us in marketing to develop disruptive strategies in the service area. I remember much the proposal of Raymond Fisk, Professor and Chair of Marketing and Logistics at the University of New Orleans in 2006, when he suggested to us the importance of considering the performance as an important tool to prepare ourselves in the attention to our students, but also with impact on the quality of care from one human being to another when it comes to interacting with commercial purposes (and in general, humans should add).

Professor Fisk argued that it was important to prepare ourselves professionally to aid a human relationship centered on the service, since the emotions that can be presented during the interaction are very variable. Human beings in and of themselves tend to stress us when we do not know another person and this is a natural reaction to protect ourselves. The idea is that, if the sales adviser knows and prepares professionally to identify the positions of fear of a client, and especially the aspects that motivated him to be in front of us, this may generate greater impact from positive messages (verbal and nonverbal).

In acting, Dr. Fisk commented, people who attend a play, do not know the mood of the person who acts (and probably does not interest them). On the contrary, they only care about achieving the goal that brought them to make a trade and leave their money with us: satisfaction.

This was precisely Dr. Fisk's main argument, the importance of preparing ourselves professionally to provide the maximum possible experience to our clients. It is not a question of deceiving them, of course not!

It is about preparing ourselves to understand them and attend them to the WOW level! Where expectations are highly overcome and emotional connections generate attitudes of loyalty. (Cardoza, 2017)

Transformative Leadership

These arguments expose the importance of generating the right organizational culture in which each and every one of the people can actually live that WOW! culture of which Cardoza speaks.

However, to generate a WOW! organizational culture requires many aligned forces, mainly a motor force that maintains the motivation of a group latent; resilient, and above all, focused on its goal of serving for love.

That strength is known as leadership, the true engine of innovation and (in the long run) evolution (objective of this chapter).

The problem of a leadership that is not centered in the love for the neighbor, is that it will usually be tempted by the corruption of the power.

Let us return once again the words of Tim Sanders:

In my opinion, if you work selflessly for your customers and collaborators to grow, that is true love. I do not know a bigger thing that you can do for a person. (Sanders, 2011)

When one person gives in to serve another, to inspire it to grow, he/she is really sharing its essence, soul, and energy. *Why would someone give someone the best of themselves if it was not for love? How is it that a person finds enough strength not to yield to the temptation of a superior position before another human being?*

Power corrupts, absolute power corrupts absolutely. Lord Acton (Pezzimenti, 2001)

The leader tends to face several challenges while working leadership positions. From the simplest to explain, such as generating a vision, the need for an organizational structure that defines goals and responsibilities, choosing and motivating the right human talent, operational efficiency, and constant growth approach; to more complex challenges to explain that are related to the understanding and correct exercise of power.

A transformative leader must inspire an organization from his or her determined hand to remove the obstacles that hold back their collaborators and a firm, but soft hand to help them rise when they stumble.

Perhaps you, dear reader, may agree with Lord Acton's phrase (Pezzimenti, 2001) and, perhaps your thoughts find justification, in the number of examples we can cite of how power has historically corrupted the judgment of several of our leaders. We may have witnessed (or worse still lived) feelings of superiority when people treat us differently by being leaders. It would not be strange that when reading these lines, one would remember those times when his office was full of people who "loved" him; however, I am sure we will also remember the loneliness of the same office when people "hated" us. In both cases, the result is unfortunately the same, the power could have corrupted it.

However, I believe that power does not corrupt. In reality, a situation of power simply exposes the bad leaders and separates them from the good ones. Power really exposes the human quality of those who exercise it.

On one occasion, a student was discussing the perverted and corrupt politics in the world and especially in our country. I shared that, for me, politics are not bad, those who are bad are the politicians that reach those positions of power corrupted by their personal and partisan interests and not by the ideals of service to their community; to their country.

The same goes for the administration. The models and theories are not bad, those that fail are the administrators. However, we must understand that failure is not bad, failure is the one word we must eradicate from our vocabulary!

The truth is, true leaders grow professionally and personally in parallel. There is no professional growth without personal growth. The true leader understands that the only way to grow professionally is by recognizing that he learns more when he looks and listens more. A true day-to-day leader challenges the human being in the mirror to be better than the day before, not forgetting his goal of serving in order to transcend.

Every day, he/she asks about how are its reactions before people, before its community. *Did his/her reactions originate from being "inside the box"? What leads to "inside the box"? How can you get out of it? What kind of emotions do you experience in situations of power?* (When you exercise it or when you exercise it on you). These and other questions could help you define your true attitude to the concept of power, which will necessarily lead you to define your true motivation for leadership.

The Sixth Element of Maslow

Usually, like human beings, companies are concerned about covering their needs according to the stage in which they are in the life cycle.

The satisfaction of each stage gives us the security to continue advancing in the pyramid of needs. We first seek the satisfaction of the basic needs. Once these are covered, we continue with the emotional ones and, later, we seek to satisfy our potential to the maximum (just like the companies do, only that the definition of basic, emotional and of performance may need a definition adjustment).

This book (and especially this chapter) speaks of a key element that defines the meaning of our existence both as human beings and as companies.

We have been talking about the 4 phases or key elements of the Stragile model:

1. Consciousness
2. Election

3. Action
4. Evolution

When we reach the 4th level, it becomes necessary to define and identify elements that allow us to guide the course in the evolution stage and that the growth or the power reached at this level does not corrupt us as individuals.

During life, we have learned that a more powerful level in terms of human existence is the level of transcendence.

For years, we have heard currents of thought that tell us that the ultimate goal of a human being is to achieve happiness. There are theories, models, and schools of thought centered on happiness.

I do not argue this. I think it is important that the human being knows how to enjoy his humanity to the fullest; however, I have always wondered if this approach is not egotistical and promotes individualism above the collective good.

From my point of view, and to be congruent with what I stated above under the love criteria of Kevin Roberts and Tim Sanders, I am more of the idea that the true feeling of happiness is reached when, out of love, we give up everything for the service of others. That is the highest level of expression of love for another human being. This is known as transcendence!

I remember the anecdote of Mother Teresa of Calcutta when she bathed a leprous brother:

Visitor: Mother, but how can you do that? I could not do it for a million dollars!

Mother Teresa of Calcutta: I would not do it for a million dollars, one bathes a leper out of love! (Van Biema. 2017)

When a human being reaches this level of communion with the universe that surrounds him, only then will he be able to identify the reasons that will transcend him in his existence and let go of superfluous happiness.

To justify this point, I found recently (just before editing this final chapter of the book), an article written by Robby Berman which proposes precisely the “missing ápex of Maslow’s Hierarchy.”, 2017.

Berman, proposes a higher level of human need, the level of transcendence. Which is defined precisely by a state of meaning and purpose of life. Just where the writing of this book began, a state of consciousness that allows us to define our purpose of life, whose course through choice and action will

Figure 23.
Source: Moreno, T. (2017), Adapted from: Berman, 2017. Maslow's Hierarchy of needs.

sooner or later take us to the next level of consciousness, called precisely evolution, since it does not, this process has no limit.

You are probably wondering at this point, *what does this have to do with the concept of evolution under the lean corporate innovation approach*?

The answer is: EVERYTHING!

Let's see: When a company has reached such a level of development and growth from living the stages of problem-solution adjustment, product market adjustment, escalation, and the last stage of lean operating efficiency, the options that are left are practically three:

1. Perpetuate its original proposal in the stage of maturity and evolve as a socially responsible company.
2. Innovate and launch a new blue ocean proposal (pioneer life cycle).
3. Continue as a cost center (until leaving the market).

The first two options comply with the Lean Corporate Innovation Model. The latter remains under the traditional premise of resistance to change.

The fourth industrial revolution presents us with totally different challenges. Automation, robotization, connectivity, the internet of things, and above all, artificial intelligence, are about to disruptively modify the world in which we have lived to this day. Moreover, there are only two options to this, either you get on the wave or you let yourself drag you with unpredictable consequences!

In the article "*The Moral Dilemmas of the Fourth Industrial Revolution*", Vinayak Dalmia and Kavi Sharma, pose the following challenges to consider in a totally different era (Dalmia, V. & Sharma, K., 2014).

The approach is interesting from an ethical point of view, but it can also be discussed under technological, penal, economic, social, academic point of view; it is a great wave!

For example, see Figure 24.

They raise different questions:

Should your self-driving car value your life over a pedestrian's? Should your Fitbit activity be used against you in a court case? Should we allow drones to become the new paparazzi? Can one patent a human gene? (Dalmia, V. & Sharma, K., 2014).

In any case the challenges and also the opportunities are immense. From the exponential growth of technology and human consciousness, the result can be devastating or hopeful:

Figure 24.
Source: Moreno, T. (2017), Adapted from: World Economic Forum, 2017.

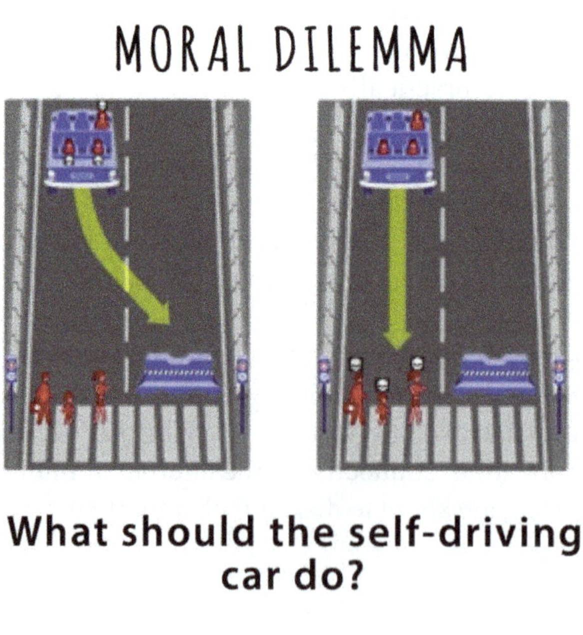

The potential benefits are huge, since everything that civilization has to offer is a product of human intelligence; we cannot predict what we might achieve when this intelligence is magnified by the tools AI may provide, but the eradication of disease and poverty are not unfathomable. Because of the great potential of AI, it is important to research how to reap its benefits while avoiding potential pitfalls. (Dalmia, V. & Sharma, K., 2014).

In any case, the answer lies in the decision that you as a leader of your company take under the argument of a collective conscience!

Integration to an Ecosystem of Innovation Beyond the Limits of an Organization

We did not want to miss the opportunity in this chapter and at the end of this book to try to go a little further. The model that Ciudad Juárez developed to move forward after one of the most devastating economic and security crises in Mexico is the basis of a collective innovation proposal (beyond the concept of corporate innovation).

The experience gained to develop the proposal of construction of a Technological Park for the Monterrey Institute of Technology and Higher Education (ITESM) in Ciudad Juárez made us understand the reality behind the expression *"in life you do not win and you lose, you win and you learn."*

After developing the executive project, submit it to the State Governor and the Secretary of Economy for approval, to process the amount corresponding to the federal government and the corresponding to the state government and to have the check literally on my desk, it was decided to return it and to thank the authorities for the support given to the project.

The reason was simple, the model of technological parks for the Tec had come to an end. It required a different kind of development, multiple research centers, technological labs, approaches, and infrastructure much leaner than a white elephant centered on the rent of offices.

Although it is fair to add that the period the city was going through represented a moment of risk for an investment of that type. Jorge Contreras, one of the city's economic and social leaders and adviser to the Tec de Monterrey in Ciudad Juárez, made it clear in a forum for consultation with the President of Mexico:

It makes no sense to make economic investments, as long as we continue to live kidnappings and continue to demand "cuotas" ("Protection" pay system used throughout Mexico by organized crime which consisted of regular payments from business owners in order to "avoid" harm.) from employers. First, we have to focus on the social and security challenge and, once recovered, we will have to take up the issue of economic growth.

A few years later, and after actively collaborating in the "*Mesa de Seguridad*" in Ciudad Juarez, this same group of businessmen invited me to make a proposal to develop a technology park for the city. My answer at the time was also clear: I do not think that the city needs a technology park and I do not think it should be a single institution that must build a scientific technological proposal that promotes economic development from education and science.

The interrogation did not wait, *how come we do not need a technology park*? Then, *what do we need*? *How do you think a proposal of this nature can be built between various institutions*? *How is it possible that you say that, if the same Tec has technology parks throughout Mexico and examples of these there are also all over the world*?

The lesson learned within the same Tec, helped me to answer:

We do not need more white elephants; we need to learn how to collaborate among institutions and generate a true ecosystem of regional innovation that integrates the current resources of community institutions and, in turn, supplements them, if required.

We need to learn to collaborate instead of competing for the same resources. We need to build a research center that will become the brain of the innovation ecosystem for the north.

We need a model that builds innovation from three elements:

1. Associativity (the ability to learn to associate with others in order to achieve joint goals)
2. The creation of public policy that promotes innovation and technology
3. The construction of technology and innovation infrastructure.

In disbelief, the answer was, "What do you need?"

Again, the response was: "*Allow me to integrate an inter-institutional team of researchers to help us integrate an executive project.*"

The model presented, of course, was based on the world concept of the triple helix (which is now said to be penta helix, considering the end user and the public and private financing schemes).

I must thank that, at the time, there was a vote of confidence in the idea. For the second meeting were already called leaders of the Technological Institute of Ciudad Juárez and the Autonomous University of Ciudad Juárez.

However, doubts remained latent. In that second meeting, I remember an expression of one of the academic leaders: "sounds utopian," he said with a tone of discredit. We ourselves thought that we were not capable of developing a concept beyond a utopia.

Figure 25.
Source: Portillo, E. (2016), Triple Helix Integration Model.

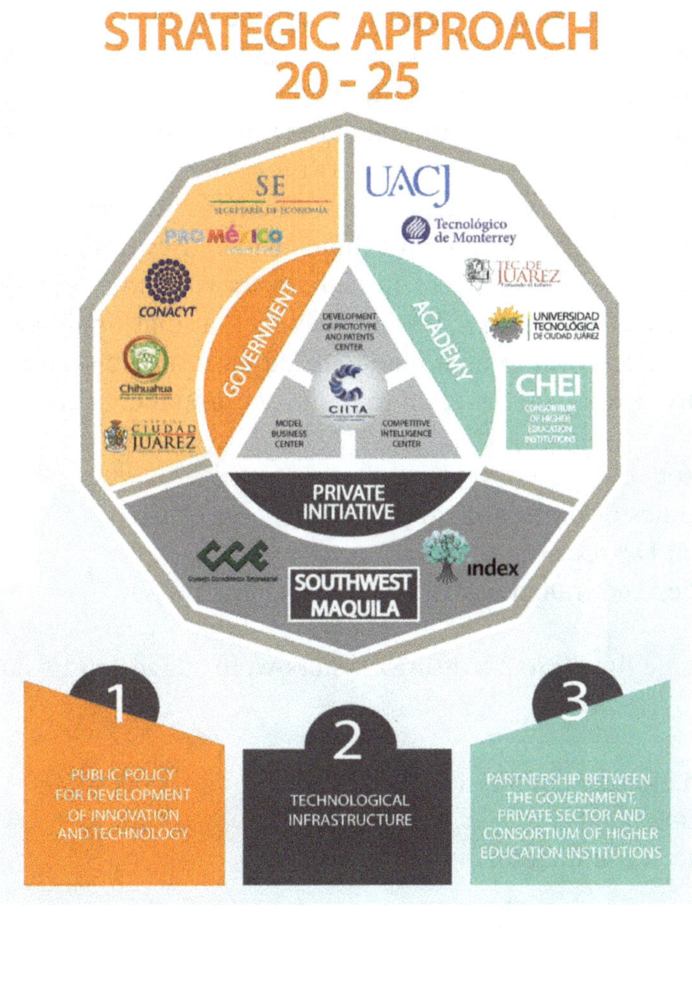

Time has proved us right. Two of the academic leaders of this project are today directors of the state government, and one more, the director of the Center for Research in Advanced Technology Integration (CIITA).

The project became the most relevant of a list of 22 initiatives that were presented to the President of the MexicanRepublic and whose Coordinator of Science, Technology, and Innovation, Dr. Elías Micha described literally as "*the type of projects that this country needs detonated.*"

The CIITA Development Model included an analysis of the different ecosystems that were currently detonating in the world. Medellin, Bogota, Barcelona, Austin, Silicon Valley, Boston, and several other examples that we were able to study.

By the way, one of the books that contributed most to the development of this concept was *Innovacities (Scheel & Pineda, 2015).*

This scientific work explains the different models of innovation worldwide, the variables of success that have been contemplated for its development, and especially, economic models of growth focused on innovation.

The analysis of this and other works related to the subject was the first step to focus the CIITA project. The work team of the project managed to gather such amount of information that the foundation of the creation of the ecosystem of innovation step of the north was already irrefutable.

In addition to this, it was also possible to analyze the proposal of the federal government for the development of the considered "*strategic sectors in the country*," and at the same time, to align them with the state development plan (for the State of Chihuahua). The curious fact, at that time, was that there was no municipal development plan focused on strategic sectors.

This analysis allowed the team to identify 4 initial clusters:

1. Automotive
2. Electronics or household appliances
3. Medical Devices
4. Logistics and transportation

At the same time, cross-cutting clusters were suggested to support the 4 pillars:

1. Information Technologies
2. Tourism
3. Advanced Manufacturing (among others).

To date, the formation and development of strategic sectors (clusters) has advanced thanks to the vision and support of the group of researchers who began this exercise (and who are supporting it from the current State Government); however, it is fair to say that, to a large extent, the clustering strategy has been developed thanks to the vision, focus, and determined commitment of the Director of PROMÉXICO in the State of Chihuahua, Dr. Carlos Yates, who has firmly insisted on the issue before the Secretariat of Economy of the Federal Government, for the integration of this initiative.

The original model is as shown in Figure 26.

In its evolution, the model has expanded from the day in which this book was published, 4 cluster pillars:

1. Medical Devices
2. Medical Tourism
3. Agroindustrial (cotton - In process)

Figure 26.
Source: Portillo, E. (2016), Advanced Technologies Research and Integration Center (CIITA).

4. Sustainable Energy (solar energy - In process)

And two cross-clusters:

1. Logistics and Foreign Trade
2. Artificial Intelligence

However, regardless of the rapid pace of growth and development of these sectoral strategies, there was a pending agenda item, which is of course included in CIITA's original model: the integration of local suppliers into productive chains, or what we call "*minicluster*."

That is to say, it would not make sense to continue developing the first level of the productive chains (based on foreign investments or related to a few families of resources of the region), if it does not include (within the strategy) a plan of integration and development of the supply which replaces the flow of world imports that now, and for more than 40 years, have reached 97% of the supply required by the industry. This is where the proposal of the Calpulli Project comes in, a term we selected at the suggestion of Mr. Manuel Miranda, for its historical connotation:

Aztec clan formed by a group of families with common ancestry that owned communal lands, an internal government and a training school for young warriors. (Oxford Dictionary, 2017)

The model of the Calpulli Project is presented in Figure 27.

The idea, put in this way, is simple: it aims to detonate a hub of innovation and human development in Ciudad Juárez (or the Paso del Norte region) from the different cluster development efforts aligned to the strategic sectors defined by the Federal and State Government, considering, of course, the inclusion and development of miniclusters in the productive chains from their incubation and acceleration in four blocks of strategic growth:

1. Competitive intelligence (roadmaps and opportunities)
2. Business Model (for each initiative to be developed)
3. Development of prototypes and patents (proposals of innovation accredited and certified to provide to different clusters).
4. Escalation or acceleration model.

Figure 27.
Source: Portillo, E., 2017, Capulli Proyect.

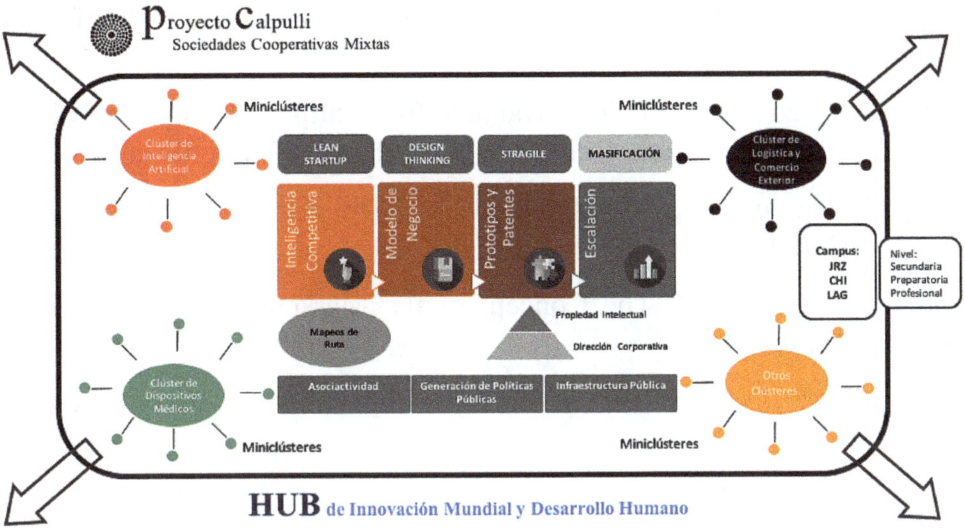

As support for this initiative, three different articles are presented in The Economist Intelligence Unit, 2016, which are directly related to this proposal.

- "Incubating innovation"
- "From design to reality"
- "Co-creation and micro factories"

In our view, the common factor of the three articles is the integration of the different actors of a true innovation ecosystem. From three different perspectives of integration:

1. Regional
2. Corporate
3. Staff

In all cases reference is made to the process of integrating efforts under an approach of innovation integration.

REFERENCES

Amaorg. (1984). *Amaorg.* Available from: https://www.ama.org/Pages/default.aspx

Berman, R. (2017). *Big Think.* Available from: http://bigthink.com/robby-berman/the-missing-apex-of-maslows-hierarchy-could-save-us-all

Berry, Parasuraman, & Zeithaml. (1990). The Service-Quality Puzzle. *Business Horizons.*

Borden, N. H. (1964). The Concept of the Marketing Mix. *Journal of Advertising Research.*

Cardoza, I. (2017). *Conversation with Motivational Talks Mexico*. Academic Press.

Cateora, P., Gilly, M., & Graham, J. (1971). *International Marketing*. The McGraw-Hill Companies, Inc.

Chan, C. (2013). *ChrisChancom.* Available from: https://chrischan.com.au/2013/06/04/innovation-4-part-recipe-for-sustaining-an-innovation-pipeline/

Csices. (2017). *AIdeclaration.* Available from: http://www.iiia.csic.es/barcelonadeclaration/

Dalmia, V., & Sharma, K. (2014). *World Economic Forum.* Available from: https://www.weforum.org/agenda/2017/02/ethics-2-0-how-the-brave-new-world-needs-a-moral-compass/

Differencebetweencom. (2015). *Differencebetweencom.* Available from: http://www.differencebetween.com/difference-between-marketspace-and-vs-marketplace/

Dixon, M., Ponomareff, L., & Turner, S. (2017). *Harvard Business Review.* Available from: https://hbr.org/2017/01/kick-ass-customer-service

Fisk, R. (2006). *Conversation with University of New Orleans.* Academic Press.

Giansante, G. (2015). *Giancarlogiansantecom.* Available from: http://giancarlogiansante.com/2015/03/12/lovemarks-la-pasion-de-los-consumidores/

Glaser, J., & Glaser, R. (2014). *The Neurochemistry of Positive Conversations*. HBR.

Kottler, P., & Keller, K. (2006). *Dirección de Marketing*. Pearson Educación.

Mccarthy, E. J. (1960). *Basic Marketing: A Managerial Approach*. Academic Press.

Oxford. (2017). *Oxford Dictionaries*. Available from: https://www.oxforddictionaries.com/

Parasuraman, A., Zeithhaml, V. A., & Berry, L. L. (1988). SERVQUAL: A multiple-item scale for measuring consumer perceptions of service quality. *Journal of Retailing*. Available from: https://www.researchgate.net/profile/Valarie_Zeithaml/publication/225083802_SERVQUAL_A_multiple-_Item_Scale_for_measuring_consumer_perceptions_of_service_quality/links/5429a4540cf27e39fa8e6531/SERVQUAL-A-multiple-Item-Scale-for-measuring-consumer-perceptions-of-service-quality.pdf

Pezzimenti, R. (2001). The Political Thought of Lord Acton: The English Catholics in the Nineteenth Century. Rome: Academic Press.

Rajagopal. (2009). *Information Communication Technologies and Globalization of Retailing Applications*. Academic Press.

Roberts, K. (2005). *Lovemarks*. Saatchi & Saatchi.

Sanders, T. (2011). *Today We Are Rich: Harnessing the Power of Total Confidence*. Tyndale House Publishers, Inc.

Scheel, C., & Pineda, L. (2015). *Innovacities*. Bogotá: UTADEO. Available from http://www.utadeo.edu.co/es/publicacion/libro/editorial/235/innovacities

Sisney, L. (2011). *Organizationalphysicscom.* Available from: http://organizationalphysics.com/2011/10/28/the-stages-of-the-execution-lifecycle/

Sisney, L. (2012). *Organizationalphysicscom.* Available from: http://organizationalphysics.com/tag/financing/

Stauffer, E. (2011). *Erich Stauffer.* Available from: http://www.erichstauffer.com/management/lean-startups

TaylorK. (2017). *Businessinsidercom.* Available from: http://nordic.businessinsider.com/the-american-retail-apocalypse-in-photos-2017-3

The Economist Intelligence Unit. (2016a). *Producers of Tomorrow*. Available from: http://producersoftomorrow.economist.com/co-creation-and-microfactories

The Economist Intelligence Unit. (2016b). *Producers of Tomorrow*. Available from: http://producersoftomorrow.economist.com/from-design-to-reality

The Economist Intelligence Unit. (2016c). *Producers of Tomorrow*. Available from: http://producersoftomorrow.economist.com/incubating-innovation-infographic

About the Authors

Jesús Enrique Portillo Pizaña has served as the Dean of the Research, Innovation, Social Development and Entrepreneurship Division of Tecnológico de Monterrey Campus Ciudad Juárez since 2011. He is also a Board member for Mesa de Seguridad y Justicia de Ciudad Juárez (Security and Justice Council for Ciudad Juárez), Fortalessa (Center for the Strengthening of Non-Governmental Organizations), The Hub of Human Innovation at El Paso Tx., and CONREDES (Industry and Academy organization that integrate Research, Innovation and Talent Development in Ciudad Juárez), Energy Cluster, Artificial Intelligence Cluster, Logistics and International Commerce Cluster, Medical Tourism Cluster, Red Juarez Emprende (entrepreneurship Cluster), among others. He received his Doctor in Business Administration with a Marketing Concentration in 2003 from Tecnológico de Monterrey Campus Ciudad de México in coordination with the Kennan-Flagler Business School at the University of North Carolina. Dr. Portillo's areas of expertise are in Strategic Marketing and Customer Development, Services Marketing, Research Methodology, Positioning and Branding, Organizational Architecture Innovation, and Entrepreneurship.

Sergio Ortiz Valdés is a scholar interested in entrepreneurship and international business. He has helped several ideas to become new companies, some of them high tech companies, he is cofounder of two technology base companies. Dr. Ortiz has started and developed the entrepreneurship research program in Tecnológico de Monterrey, designing, organizing and implementing a special PhD program focusing in entrepreneurship and entrepreneurship education. Dr. Ortiz has been part of the different Mexican delegations representing Mexico in international affairs, like the North America Committee, Mexico-Europe bi-National committee, China-Mexico bi-National Committee, Mexico China high level group. Dr. Sergio Ortiz has participated in different consultancy projects with several companies in fields related with innovation. Dr. Ortiz has been Dean of Business School in Monterrey Tech, Mexico State Campus, and Director of the Technological Entrepreneurship Program in Monterrey Campus. He also worked as a Financial Analyst and Economic Analyst for the Mexican Central Bank. Dr. Ortiz is economist with MBA in Tecnológico de Monterrey, Master of International Management in Thunderbird University, and holds a PhD in Administrative Sciences in the University Antonio de Nebrija, Spain. His actual position is Academic Director for Entrepreneurship in Tecnológico de Monterrey Presidency, responsible for reinventing entrepreneurship and innovation education in the 31 campuses of Tecnológico de Monterrey.

Luis Miguel Beristain Hernández is a student, practitioner and teacher of entrepreneurship, innovation and technology transfer. He has been co-founder of more than twenty companies throughout his career, supported more than six hundred entrepreneurs to realize their business vision and *angel investor* in twenty projects. His experience includes the chemical, textile, health services, informatics and, in recent years, medical engineering and biotechnology applied to health sectors. He has served as a leading consultant in more than 18 companies, specializing in "rescuing companies in crisis." He has various publications in national and international refereed journals, as well as the publication of several book chapters, always with the topics of product innovation, organizational innovation, intellectual property, technology transfer, evaluation and technology valuation. He was recently appointed Director of the Eugenio Garza Lagüera Entrepreneurship Institute of the South Zone of the Tecnológico de Monterrey, an institution where for more than twenty-five years he has worked as a professor and occupied various positions in the Mexico City Region as Director of the Entrepreneurship Institute; at the Mexico City Campus as Director of the School of Medicine and Life Sciences, Director of the Science in Life Technology Park, Director of the Entrepreneurship Division, Director of the Division of Health Sciences, Director of Academic Support, Director of Liaison, and Director of Administration and Finance. He was also Division Director and General Director of the Central Campus of Veracruz. He holds a Ph.D. in Administrative Sciences from the Tecnológico de Monterrey, a Master's in Business Administration from the Pan-American Institute of Senior Business Management and a Bachelor in International Relations from the Universidad de las Américas-Puebla.

Index

21st century 133, 142

A

Altair 8800 204, 243
alternative solutions 44, 56

B

business system 150, 209, 211, 216-217, 224, 234-238, 240, 242-243, 246

D

DNA Definition 42

E

Elektra 180-181, 224-225, 228, 231, 239, 241

G

gestating companies 150-151, 178

H

human talent 4, 19, 27, 32, 43, 47, 65-66, 77, 133, 146-148, 247, 252, 259, 278

I

incubation stage 19, 42, 66
Innoduct 165, 178
innovation 4-11, 17, 26, 28-30, 34, 43-44, 46-47, 49, 57, 67-68, 71-73, 75, 78-79, 83, 89, 91, 96-97, 99-101, 103-104, 107, 109-119, 121, 124-125, 127-130, 133-135, 138, 140-148, 150-152, 155-156, 158, 160, 165, 167, 169, 176, 178-179, 181-182, 184-185, 189, 192, 195, 197, 204, 206-208, 212, 237, 243, 247-249, 253, 256-259, 261-262, 277, 281, 283-284, 286, 288-289
Ivy Leage 134

K

Key Aspects 100, 132

L

life cycle 10, 26, 61, 67, 71, 100, 154, 164, 176, 206, 247, 253-254, 261-263, 279

M

Maquiladora Industry 4, 42
market 7, 10-13, 26, 28-29, 34, 42, 45, 61-62, 66-72, 74-76, 78-83, 85-86, 88-

89, 91, 93, 98-99, 102-104, 111-116, 118-119, 121, 123-125, 127-129, 137, 150-155, 157, 161, 163, 165, 168-169, 171, 174-176, 180-183, 186-189, 192, 195, 198, 203-204, 206-212, 215, 218-225, 229-236, 239-240, 243, 246, 248, 251-252, 254-256, 260, 263, 270, 281
market economy 211, 218-219, 246
Mega Trends 42

P

product 4, 9, 12-13, 62, 66, 68-72, 74-76, 78-83, 86, 91, 102-103, 111, 117, 124-127, 145, 150, 152-155, 157, 164, 166, 168, 170-172, 174-176, 178, 180-183, 187-195, 197-198, 203-207, 209, 212-213, 219-220, 223, 229, 231-234, 238-241, 243, 252, 254-256, 259-265, 270, 274, 281, 283
prototype 117, 150, 158, 163, 167-176, 180, 184, 189-190, 193, 198

R

Red Ocean 30, 42

S

self-efficacy 111, 132, 135
Silos 42
Sine Qua Non 151, 178
Spinoff 42
step 1, 3, 9, 13, 21, 25, 29, 36, 43-45, 47, 49, 56-57, 59, 61, 65-66, 71, 77, 91, 98-99, 103-105, 107, 109, 121, 126, 134, 141, 150, 186, 197, 204, 230, 239, 247, 252, 269, 286
Stragile model 8, 11, 17, 19, 43, 47, 49-50, 61, 97, 100, 279
strategic adequacy 36, 45, 70

U

user 86, 116-117, 127, 143, 150, 155, 157, 161-163, 169-170, 172, 174, 178, 189-190, 285

W

white elephant 6, 42, 283

www.ingramcontent.com/pod-product-compliance
Lightning Source LLC
LaVergne TN
LVHW081524300926
849161LV00006B/1291